Cambridge Elements ≡

Elements in the Philosophy of Biology
edited by
Grant Ramsey
KU Leuven
Michael Ruse
Florida State University

INHERITANCE SYSTEMS AND THE EXTENDED EVOLUTIONARY SYNTHESIS

Eva Jablonka

Marion J. Lamb
*Cohn Institute for the History and Philosophy
of Science and Ideas
Tel Aviv University, Israel*

CAMBRIDGE
UNIVERSITY PRESS

CAMBRIDGE
UNIVERSITY PRESS

University Printing House, Cambridge CB2 8BS, United Kingdom

One Liberty Plaza, 20th Floor, New York, NY 10006, USA

477 Williamstown Road, Port Melbourne, VIC 3207, Australia

314–321, 3rd Floor, Plot 3, Splendor Forum, Jasola District Centre, New Delhi – 110025, India

79 Anson Road, #06–04/06, Singapore 079906

Cambridge University Press is part of the University of Cambridge.

It furthers the University's mission by disseminating knowledge in the pursuit of education, learning, and research at the highest international levels of excellence.

www.cambridge.org
Information on this title: www.cambridge.org/9781108716024
DOI: 10.1017/9781108685412

First published 2020

A catalogue record for this publication is available from the British Library.

ISBN 978-1-108-71602-4 Paperback
ISSN 2515-1126 (online)
ISSN 2515-1118 (print)

Inheritance Systems and the Extended Evolutionary Synthesis

Elements in the Philosophy of Biology

DOI: 10.1017/ 9781108685412
First published online: June 2020

Eva Jablonka and Marion J. Lamb
Tel Aviv University
Author for correspondence: Eva Jablonka jablonka@tauex.tau.ac.il

Abstract: Current knowledge of the genetic, epigenetic, behavioural and symbolic systems of inheritance requires a revision and extension of the mid-twentieth-century, gene-based 'Modern Synthesis' version of Darwinian evolutionary theory. We present the case for this by first outlining the history that led to the neo-Darwinian view of evolution. In the second section we describe and compare different types of inheritance, and in the third we discuss the implications of a broad view of heredity for various aspects of evolutionary theory. We end with an examination of the philosophical and conceptual ramifications of evolutionary thinking that incorporates multiple inheritance systems.

Keywords: heredity, development, epigenetics, plasticity, Modern Synthesis

ISBNs: 9781108716024 (PB), 9781108685412 (OC)
ISSNs: 2515-1126 (online), 2515-1118 (print)

Contents

1 The Modern Synthesis: a Neo-Darwinian, Genotypic View of Heredity and Evolution

Since around the turn of the century, the idea that mainstream evolutionary theory needs substantial revision has been the subject of vigorous and sometimes vitriolic debate. Some evolutionary biologists maintain that any attempt to radically revise the present view is not only unnecessary and misguided but also dangerous. Any questioning of evolutionary ideas plays into the hands of creationists, they claim.

We are among those who believe that a change in the way we think about evolution is overdue, and in this Element we look at the debate from the perspective of heredity. We argue that what has been learned about genetic, epigenetic, behavioural and symbolic systems of inheritance in the past 50 years requires a substantial revision and extension of the mid-twentieth-century, gene-based 'Modern Synthesis' (MS) version of evolutionary theory. We need to return to an earlier, development- and organism-oriented view. As Lenin said, the evolution of a science is 'a development that repeats, as it were, the stages already passed, but repeats them in a different way, on a higher plane ... a development, so to speak, in spirals, not in a straight line' (1914/1930, p. 14). We see the version of evolutionary theory that is being advocated today, commonly called the 'Extended Evolutionary Synthesis' (EES), as an updated version of the early twentieth-century organicists' evolutionary view on a higher plane, growing from, yet also challenging, the Modern Synthesis.

We start here with a historical overview, outlining the origins of the MS and the nature of the challenge to it. In Section 2 we describe the different types of heritable variations that are interacting inputs into the development of phenotypes. Our emphasis will be on epigenetic inheritance, which is found in all forms of life and is the molecular basis of non-genetic inheritance. The evolutionary implications of multiple inheritance systems are discussed in Section 3, and in the final section we explore the implications of this expanded view of heredity and evolution for the philosophy and sociology of biology.

1.1 The Modern Synthesis

In 1942, in the middle of World War II, a book with the title *Evolution, the Modern Synthesis* was published in London. Its author was Julian Huxley. Like his grandfather, the ardent Darwinian Thomas Henry Huxley, Julian Huxley was a zoologist, and as a young man published scholarly work on bird behaviour, experimental embryology and evolution, but he was also a gifted popularizer of science. Through books and articles, broadcasts and films, he introduced lay people to the excitement of the new ideas that were emerging in biology. One of his most successful projects was *The Science of Life*, a three-volume book that was originally issued in fortnightly parts. This popular work, completed in 1930, was co-authored by the novelist and essayist H. G. Wells and his zoologist son G. P. Wells. It presented a unified view of biology, with evolution as a central theme. Huxley's 1942 book, *Evolution, the Modern Synthesis*, aimed mainly at professional biologists, gave a more focused and academic state-of-the-art account of evolutionary ideas. The title was appropriate – it was modern, and it was a synthesis.

Today, probably very few biologists will have read any of Julian Huxley's writings, but all are familiar with the expression 'the Modern Synthesis' (which we refer to as the MS). This phrase, along with the variants 'the evolutionary synthesis' and 'the synthetic theory of evolution', has entered the lexicon of biology as a summary-term for a view of evolution that was developed in the second quarter of the twentieth century. It had its roots in books written by geneticist Theodosius Dobzhansky, zoologist Ernst Mayr, palaeontologist George Gaylord Simpson and botanist G. Ledyard Stebbins, as well as Julian Huxley and a few other authors. These books, published between 1937 and 1950, vary in the emphasis they put on different aspects of biology, but all acknowledged the significance of the belated marriage of Darwin's theory of natural selection and Mendelian genetics (reviewed in Bowler, 2003; Smocovitis, 1996).

In the late nineteenth and early twentieth centuries, the popularity of Darwin's theory of gradual adaptive evolution through natural selection slumped, but it was revived during the 1920s and 1930s through the work of the population geneticists Ronald Fisher and J. B. S. Haldane in Great Britain and Sewall Wright in the USA (Provine, 2001). They showed mathematically how genetic differences among individuals in a population could lead to adaptation. There was general agreement that evolution usually involves a gradual, cumulative change in gene frequencies in populations, brought about by selection acting on the variation among individuals that results from random gene mutation and recombination. This became the orthodox 'Modern Synthesis' view that dominated evolutionary thinking for most of the rest of the century, a view that hardened as the century progressed.

Not only could gradual selection acting on the abundant variation produced by small random mutations explain adaptedness within populations (microevolution), but the same process could explain the origin, multiplication and diversification of species and higher taxonomic categories (macroevolution). Mayr, in particular, maintained that macroevolution usually occurred through the gradual genetic restructuring of populations during long periods in which they are biologically (reproductively) isolated from other populations. His summary of the origins and major assumptions of the MS view was:

> It was in these years [1936–1947] that biologists of the most diverse subdivisions of evolutionary biology and from various countries accepted two major conclusions: (1) that evolution is gradual, being explicatory in terms of small genetic changes and recombination and in terms of the ordering of this genetic variation by natural selection; and (2) that by introducing the population concept, by considering species as reproductively isolated aggregates of populations, and by analyzing the effect of ecological factors (niche occupation, competition, adaptive radiation) on diversity and on the origin of higher taxa, one can explain all evolutionary phenomena in a manner that is consistent both with the known genetic mechanisms and with the observational evidence of the naturalists. Julian Huxley (1942) designated the achievement of consensus on these points as *the evolutionary synthesis*. It required that the naturalists abandon their belief in soft inheritance[1] and that the experimentalists give up typological thinking and be willing to incorporate the origin of diversity in their research program. It led to a decline of the concept of 'mutation pressure,' and its replacement by a heightened confidence in the powers of natural selection, combined with a new realization of the immensity of genetic variation in natural populations. (Mayr, 1982, p. 567, his italics)

Heredity was thus clearly identified with genetics. Dobzhansky, another of the founding fathers of the MS, defined biological heredity not only in terms of genes but in terms of self-serving genes:

> Heredity is, in the last analysis, self-reproduction. The units of heredity, and hence of self-reproduction, are corpuscles of macromolecular dimensions, called genes. The chief, if not the only, function of every gene is to build a copy of itself out of the food materials; the organism, in a sense, is a by-product of this process of gene self-synthesis. (Dobzhansky, 1958, p. 21).

1.2 The MS Notion of Heredity

The development of the genocentric notion of heredity, which in the 1970s became integral to Dawkins' 'selfish gene' view of evolution, was strongly influenced by August Weismann's neo-Darwinian synthesis and Wilhelm Johannsen's genotype–phenotype distinction. Weismann (1889) combined Darwin's view of evolution with his own highly speculative theory of heredity

---

[1] Soft inheritance was the term Mayr used 'to designate the belief in a gradual change of the genetic material itself, either by use and disuse, or by some internal progressive tendencies, or through the direct effect of the environment. ... All theories of soft inheritance deny the complete constancy of the genetic material that we now know to exist' (Mayr, 1980, p. 15). For a brief history of attitudes to soft inheritance, see Jablonka and Lamb (2011).

and development, which involved sets of hierarchically organized hereditary units located in the chromosomes of cell nuclei. According to Weismann, the architectural organization of the units meant that when somatic cells divide, two daughter cells can inherit different sets of units; during development and differentiation, this organization also causes units to move out of the nucleus and take control of the cell's activities. Consequently, during cell division and differentiation, somatic cell nuclei progressively lose hereditary units and their capacity to form all parts of the body. Only germplasm, found mainly in the germline cells that give rise to gametes, retains complete sets of hereditary units. Therefore, any induced or acquired changes that occur in somatic cells during the developmental history of individuals cannot be passed to future generations. This belief became known as 'Weismann's doctrine of the independence and continuity of the germplasm'. 'Weismann's barrier' explained why the inheritance of acquired characters (inappropriately known as 'Lamarckism') is impossible.[2]

Weismann's view of evolution, with its extreme selectionism and rejection of Lamarckism, was soon dubbed 'ultra-Darwinism' or 'neo-Darwinism' to distinguish it from Darwin's more pluralistic approach. It was widely discussed, but growing knowledge about the behaviour of cells and chromosomes, and the emphasis on experimentation, particularly after the recognition in 1900 of the significance of Mendel's work, led to the rejection of many of Weismann's ideas. His doctrine of the continuity of the germline survived, however.

Wilhelm Johannsen, a Danish botanist, was a critic of the 'morphologico-phantastical speculations of the Weismann school' (1911, p. 133). His work on 'pure lines' – populations of beans and other self-fertilizing plants in which all individuals are descendants of a single parent – led Johannsen to what he called the 'genotype conception of heredity'. Pure lines do not respond to selection, he reasoned, because all individuals are 'genotypically' identical; their 'phenotypes' – their outward appearances – differ, but these differences are the result of environmental effects that are not inherited. Any new selectable variations appearing in pure lines are the outcome of alterations in the constituents of the genotype – of mutations induced by changes in conditions.

Unlike Weismann, Johannsen was unwilling to speculate about the material basis of heredity. In order to distance himself from the mechanistic and deterministic entities that Weismann and others had suggested, he called his abstract

---

[2] The inheritance of acquired characters was an almost universal belief among natural philosophers until the late nineteenth century, and was not specific to Lamarck. Weismann's view of the inheritance of acquired characters was more nuanced than is usually assumed. He accepted that the environment could act directly on hereditary elements in the germ cells and that selection within these cells plays a significant role in evolution (Weissman, 2011).

elements 'genes'. He then defined a genotype as 'the sum total of all the "genes" in a gamete or in a zygote' (Johannsen, 1911, pp. 132–3) and concluded: 'Heredity may then be defined as *the presence of identical genes in ancestors and descendants*' (p. 159, his italics). Like Weismann, Johannsen concluded that there is a one-way route between an organism's hereditary endowment and its physical characteristics. Acquired characters cannot be inherited, because there is no mechanism through which developmental modifications can be transmitted to future generations. This was a belief that helped shape the genetic view of heredity that was soon to be incorporated into the MS.

As Mendelian genetics came to dominate thinking about heredity, the meaning of Johannsen's terms shifted. 'Genes' were increasingly thought of as real physical particles located on chromosomes rather than as abstract entities. The meaning of the terms 'phenotype' and 'genotype' also changed, as increasingly they were applied to a single character and the pair of genes (alleles) associated with it as well as the total genetic constitution and appearance. 'Mutation' came to be used for random changes in the material make-up of a gene and for the processes bringing about such changes.

Because they usually studied the inheritance of distinct, alternative, discontinuous characters, many of the early Mendelians thought that evolution occurred in mutational leaps. Other biologists, particularly the naturalists, defended Darwin's gradualism, insisting that the large differences the Mendelians studied were of little significance in evolution. However, by the 1930s, mutation theories had faded in importance for all but a few diehards like German-American geneticist Richard Goldschmidt, who insisted that 'systemic' mutations, which involved a repatterning of the genome, as well as occasional regulatory mutations with large effects, were behind many macro-evolutionary changes. For the quantitative and population geneticists whose mathematical analyses contributed so much to the MS, mutations were the small, rare and random changes in genes that produced new alleles. Evolutionary change had more to do with selection, population size, the system of mating, migration and gene interactions than with rare mutations.

1.3 The MS as a Unifier of Biology

The MS was seen by some of its architects as far more than an updated, Mendelized and mathematized version of neo-Darwinism. This is clear from the title of a book edited by Mayr and Provine (1980), *The Evolutionary Synthesis: Perspectives on the Unification of Biology*, which was based on a conference held in 1974 to discuss the construction of the evolutionary synthesis. As Smocovitis (1996) noted, this title reflects the ambition of the

MS to unify the many branches of biology within a coherent, truly scientific, theoretical framework. Smocovitis suggests that this ambition was comparable to that of the logical positivists of the Vienna Circle, who during the first third of the twentieth century pursued the Enlightenment ideal of unifying all science on the basis of physics. However, evolutionary biologists were not fully satisfied with this. Early twentieth-century philosopher–biologists such as J. S. Haldane (father of geneticist J. B. S. Haldane) and Joseph Henry Woodger also asserted the autonomy of biology. Though based on physics, biology cannot be reduced to physics. It was through evolutionary biology, Huxley, Mayr and others claimed, that the different branches of biology could be unified as an autonomous science.

The MS has always been more of an interpretative framework than a well-defined theory. It has been variously described as 'a harmonization of ideas', 'a coherent set of beliefs', 'a disciplinary matrix' and 'a treaty'. But however it is described and whatever its aims, population genetics has always been at its core. By building on the concepts of population and quantitative genetics, biologists coming from systematics, biogeography, palaeontology, comparative anatomy and ecology could interpret their findings within a neo-Darwinian framework. Laboratory experiments and studies of natural populations of fruitflies, snails, moths and butterflies confirmed (more or less) some of the predictions of the geneticists' models. Moreover, as genetics became more molecular, the flexibility of the MS framework allowed it to be updated. Genes were identified with sequences of DNA; heredity was seen in terms of the replication of DNA sequences; mutations were the outcome of unrepaired damage to DNA or errors occurring during its replication; and development could be described in terms of gene expression and its regulation. According to Crick's 'central dogma', information in DNA sequences is transcribed into RNA, and RNA is translated into proteins, but it cannot flow in the opposite direction from protein to RNA or DNA. The central dogma was seen as a molecular version of Weismann's doctrine and Johannsen's genotype–phenotype distinction – information flows from germline to soma, genotype to phenotype, or DNA to protein, but never in the opposite direction. Consequently, the inheritance of acquired characters is impossible. According to John Maynard Smith, a leading British evolutionary biologist, 'The greatest virtue of the central dogma is that it makes it clear what a Lamarckist must do – he must disprove the dogma' (Maynard Smith, 1966, p. 66). Of course, this would be the case only if DNA could not be directly changed by environmental inputs, and if other inheritance systems, which can transmit developmentally induced variations that are independent of variations in DNA, did not exist. Both of these assumptions can be challenged (see Sections 1.4 and 2).

Today's defenders of the MS constantly claim that new observations and ideas about evolution do not require a change in the fundamental assumptions of the MS. It has been able to absorb the discovery that there is far more DNA variation than had been expected; that many alleles seem to be neutral; and that some DNA seems to be parasitic. By incorporating the notions of kin selection and inclusive fitness, an explanation was provided for aspects of behaviour, such as altruism and cooperation, that had at first been difficult to explain within the MS framework. When Gould and others argued that because the palaeontological record of some groups showed long periods of stasis punctuated by periods of rapid evolution and divergence, something other than gradual adaptive change was occurring, MS adherents responded by saying that no Goldschmidtian systemic mutations or any other special macroevolutionary processes need be invoked. Immense periods of time are involved, they argued, even in periods of rapid change, and most change probably occurred in small populations that were unlikely to have left traces in the fossil record. Genes, gradual selection and random variation can explain everything, they continue to insist. Consequently, current textbooks of evolutionary biology use the same framework as that constructed in the 1930s and 1940s.[3] Population geneticists such as Brian Charlesworth and his colleagues, some of today's standard-bearers for the MS view of evolution, use a definition that is almost identical to that given by Dobzhansky 50 years ago:

> The core tenet of the MS is that adaptive evolution is due to natural selection acting on heritable variability that originates through accidental changes in the genetic material. Such mutations are random in the sense that they arise without reference to their advantages or disadvantages .... Because this viewpoint asserts that natural selection acts to increase the frequencies of advantageous variants within populations, it is often referred to as neo-Darwinism. (Charlesworth, Barton and Charlesworth, 2017, p. 1).

1.4 What the MS Excluded

The problem with the MS, particularly after it hardened in the 1960s and 1970s, is that it ruled out or marginalized certain views of heredity and evolution. Mayr was very explicit about which ideas were excluded, or, as he put it, 'misconceptions' that were 'refuted':

---

[3] A comparison of 10 current textbooks of evolutionary biology has shown that only gene selection, genetic drift, gene flow and gene mutation are seen as major evolutionary processes. Inclusive inheritance, plasticity and niche construction are either totally ignored or get only modest treatment (Laland et al., 2015).

> In the short run, it was perhaps the refutation of a number of misconceptions that had the greatest impact on evolutionary biology. This includes soft inheritance, saltationism, evolutionary essentialism, and autogenetic theories. The synthesis emphatically confirmed the overwhelming importance of natural selection, of gradualism, of the dual nature of evolution (adaptation and diversification), of the populational structure of species, of the evolutionary role of species, and of hard inheritance. (Mayr, 1982, p. 570)

Mayr and the other subscribers to the MS thus excluded from it certain theories (Lamarckism, orthogenesis), outlooks (essentialism) and mechanisms (soft inheritance). There was no room in the MS for any non-gradual, goal-directed or internally driven processes, and no room for the inheritance of acquired characters or any other type of 'soft inheritance'. Darwinism was redefined: 'The term "Darwinism" in the following discussions refers to the theory that selection is the only direction-giving factor in evolution' (Mayr, 1980, p. 3). This was certainly not Darwin's Darwinism – it was a version of neo-Darwinism, but labelling this view as 'Darwinism' undoubtedly endowed it with more authority.

The MS view of non-human evolution recognizes only the genetic inheritance system. It is assumed that the only mechanisms underlying the patterns and outcomes of evolution that are studied in disciplines such as embryology, systematics and palaeontology are those that lead to changes in gene frequencies in populations. Although the views held by the early promoters of the MS were complex and changed with time,[4] a genic perspective, a commitment to hard inheritance, and a bottom-up evolutionary population genetic analysis were common to them all and persist today. This is the basis of what we see as the major positive (roman) and negative (italicized) overlapping assumptions of the MS (based on Jablonka and Lamb, 2010):

1. Heredity occurs through the transmission of germline genes, and hereditary variation is caused by variation in DNA base sequence. *There are no inherited non-DNA variations that cannot be explained in terms of DNA variations. The genetic–DNA system is the only source of heritable variation.*

2. New hereditary variation is the consequence of (i) the recombination of pre-existing alleles that are generated during sexual processes and (ii) mutations – the result of accidental changes in DNA. *Hereditary variation is not affected by the developmental history of the individual. There is no soft inheritance.*

---

[4] For discussions of differences and changes in evolutionary thinking in the twentieth century, see Jablonka and Lamb, 2011; Lamb, 2011; Mayr and Provine, 1980.

3. The ultimate unit of selection is the gene. Although genes interact, and the interactions are often non-linear, the additive fitness-effects of single genes drive evolution by natural selection. *The phenotype generated by a genetic–developmental network is not heritable and cannot be a unit of evolution.*

4. Developmental canalization and plasticity are products of evolution and affect evolutionary change only by constraining the range of variation on which selection can act. *Developmental plasticity does not drive coordinated developmental changes that can guide adaptive evolution, speciation, evolutionary trends and evolutionary rates.*

5. Evolution is typically gradual, because only variations with small effects are likely to be beneficial. *Variations with large effects are almost always lethal.*

6. Large-scale evolutionary innovations that involve coordination among parts and processes are the results of the accumulation of gene mutations with small effects. *Fundamental physico-chemical processes and processes of developmental accommodation are not a primary generative cause of morphological and physiological innovations. They are merely boundary conditions.*

7. Macroevolution is the outcome of cumulative microevolution. *With few exceptions, macroevolution does not require any extra factors beyond those operating during microevolution.*

8. Conspecifics in groups interact and may coevolve with each other, with their symbionts and parasites, and with their abiotic environment. *The evolutionary effects of transferred ecological legacies that result from these interactions are relatively unimportant.*

9. The individual is a major target of selection. *Somatic selection within the individual, selection among groups, lineage selection, species selection and different forms of community selection occur but are not of major importance.*

10. Evolutionary change occurs mainly during vertical descent from a single common ancestor. *Forms of genetic exchange and sharing such as gene transfer and hybridization have minor significance; they do not alter the basic branching structure of phylogenetic divergence.*

We have no argument with the MS's positive assumptions, but we reject the negative (italicized) ones. For example, we agree that heritable variation, especially at the molecular level, is often blind to function; what we reject is the assumption that there are no developmentally acquired heritable variations (not necessarily genotypic). Similarly, although heritable variations can have small effects, not infrequently they have very large effects, which can be the basis of saltationary macroevolutionary changes.

1.5 Marginalized Ideas: Waddington's Developmental– Evolutionary Synthesis

One topic that is generally acknowledged not to have been integrated into the MS is developmental biology. This is usually said to be the result of the lack of an adequate theoretical framework and of any empirical evidence that it is important. Yet, in fact, both theory and empirical evidence were in place during the consolidation period of the MS between the 1950s and the 1980s. In the USSR a synthesis between heredity, development and ecology had been made by Ivan Schmalhausen in the late 1930s, but apart from a 1949 translation of one of his books (which was edited by Dobzhansky), it was largely unknown in the West. In Britain, beginning in the early 1940s, Conrad Hal Waddington forged a wide-ranging development-oriented evolutionary synthesis, which, although accessible to evolutionary biologists and initially discussed favourably in Britain, had little impact in the USA, where the MS was hardening. With the rise of molecular biology in the 1960s and 1970s, it came to be seen as out of date, and generated little interest until the last decade of the century. Today, developmental–evolutionary ideas of the type that Waddington put forward are at the core of the EES.

Waddington was an embryologist who turned to genetics to help him solve developmental and evolutionary problems. He began working as a biologist in the 1930s and was strongly influenced by Whitehead's process philosophy and by a group of Cambridge biologists and philosophers who formed the Theoretical Biology Club. The group's 'Biotheoretical Gatherings' took place from 1932 until the outbreak of World War II. One of their main concerns was with the still ongoing arguments between the vitalists and the mechanists, and they attempted to reframe the holistic attitude of the vitalists in materialistic, non-metaphysical terms. They adopted what they called a 'third way', an approach discussed earlier by J. S. Haldane, Lancelot Hogben and others who focused on the dynamic organization and order of living organisms in space and time, and on emergent properties that arise through interactions among parts, with parts determining wholes and wholes determining parts. Process, whole-ness, integration and networks of interactions were central to this view (Peterson, 2016).

Waddington developed his biological outlook within this framework. From the 1930s onwards, he was actively experimenting in and theorizing about embryol-ogy, genetics and evolution. He worked in Hans Spemann's embryology lab on the 'organizer'; he wrote a theoretical paper with J. B. S. Haldane on inbreeding and linkage; and he did research in Thomas Hunt Morgan's lab on genes affecting the development of the fruitfly *Drosophila*. In the 1940s and 1950s, at the time when the MS was being consolidated, he wrote a series of papers and books that presented a broad synthesis between genetics, evolution and development, arguing that selection of chance mutation is insufficient for understanding adaptive evolution.[5] In one short paper in 1942, provocatively titled 'Canalization of development and the inheritance of acquired characters', he argued that acquired adaptations can become inherited characters, and used the callosities on the sternum and rump of the ostrich, which are present at hatching, as an illustration. He suggested that selection for the developmental response to an environmental condition (in the case of the ancestral ostrich, it would be the development of protective hardened skin in response to squatting on a rough surface) has to be incorporated into evolutionary theory to make sense of the origin of adaptations.

To capture the idea he was promoting, Waddington suggested a new term, *epigenetics*, to describe the relationship between genetics and development. He later wrote:

> Some years ago (1947) I introduced the word 'epigenetics', derived from the Aristotelian word 'epigenesis', which had more or less passed into disuse, as a suitable name for the branch of biology which studies the causal interactions between genes and their products which bring the phenotype into being. (Waddington, [1968]1975, p. 218)

To understand evolutionary changes in phenotype, Waddington argued, you need to understand the 'epigenotype' and 'epigenetic processes'. Fundamental to his way of thinking was the idea that development is canalized, or buffered: tissues are normally nerve, muscle, kidneys, and so on, not something in between; in the wild, although individuals in a population differ in their genetic make-up and the conditions in which they live, they resemble each other very closely. According to Waddington, the constancy of the end results of development is the outcome of the activities and interactions of a large suite of genes that have been adjusted by natural selection to bring about one or a few definite end results regardless of small variations in conditions and genotypes.

Canalization breaks down in some extreme environments and with some rare mutations, so abnormal variant phenotypes are produced. Using fruitflies, Waddington showed that if the environment in which they develop is

---

[5] Waddington's ideas are described in *The Strategy of the Genes* (1957) and articles included in *The Evolution of an Evolutionist* (1975). We use the latter source for some citations; the publication date of the original paper is given in square brackets. For good philosophical and historical accounts that contextualize Waddington's ideas, see Baedke, 2018; Nicholson and Dupré, 2018; Peterson, 2016.

manipulated in ways that produce an abnormal character in a few flies, then selective breeding over several generations (10–20) from those individuals could establish a lineage in which most flies showed the abnormal phenotype even in normal conditions. In Waddington's terminology, the trait is 'genetically assimilated'; it has become canalized. Although this looks 'Lamarckian', in that the acquired character has become an inherited one, it was brought about through natural selection having changed the frequencies of genes in complex interacting networks. What was central to this analysis was that it started from an altered phenotype, usually initiated by changing the environment during development. Selection of gene networks that stabilized the new developmental path then followed.

Waddington recognized that development is intertwined with ecology and behaviour and argued strongly that these aspects of biology must be brought into mainstream evolutionary theory. He presented his system in a diagram (Figure 1), which shows that an animal chooses its niche and that that niche influences the development of the phenotype. Acquired morphological and behavioural features can be perpetuated by descendants because they are reconstructed developmentally in the environmental niche their ancestors bequeathed to them.

Although Waddington recognized that '[a] few animals can pass on a meagre amount of information to their offspring by other methods' ([1961]1975, p. 272), he paid little attention to non-genetic transmission. He mentions substances transferred from mothers to offspring in the cytoplasm of the egg and through the placenta and milk in mammals, tentatively and somewhat awkwardly describing them as 'mechanisms for passing into the offspring certain of the results of the activities of epigenetic gene-groupings in her body. ...In the subhuman world, however, the para-genetic mechanisms which can be utilized for transmitting to the next generation the results of epigenetic interactions between genes are only rather slightly developed' ([1961]1975, p. 290). Waddington also mentions song mimicry in birds as a way in which animals can transmit information, but in general he downplayed the significance of non-genetic inheritance through social learning in animals. Man alone, he insisted, 'has developed a sociogenetic or psychosocial mechanism of evolution which overlies, and often overrides, the biological mechanism depending solely on genes' ([1961]1975, p. 272).

Other biologists gave far more weight to the evolutionary importance of behavioural transmission in non-human animals. At the end of the nineteenth century, just before the rediscovery of Mendel's ideas, James Mark Baldwin, Henry Fairfield Osborne and Conwy Lloyd Morgan independently developed an evolutionary scenario which they called 'organic selection' but is now

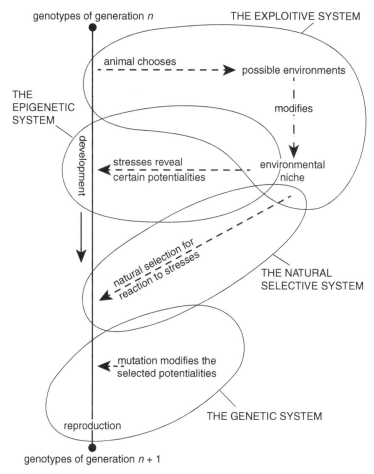

Figure 1 Waddington's evolutionary system. (Waddington, 1960, p. 401; reproduced with permission of the University of Chicago Press)

known as the 'Baldwin effect' (discussed in Hardy, 1965; Weber and Depew, 2003). Their basic argument was that plasticity, especially behavioural plasticity, allows animals to adapt to new environments during their own lifetime; then, through natural selection in the new environment over many generations, hereditary variations that improve or replace the plastic response spread. As Lloyd Morgan put it, 'Thus plastic modification leads and germinal variation follows; the one paves the way for the other' (cited in Hardy, 1965, p. 167). The scheme proposed by the inventors of the organic selection concept included an important role for 'social heredity' or 'traditions' through which young animals learn how to behave from their parents and other adults. Although in his 1942 book Julian Huxley recognized the likely importance of organic selection, and in the 1950s and 1960s Alister Hardy in the UK stressed how patterns of behaviour that are transmitted through social learning could drive the genetic evolution of many animal groups, the influential American founding fathers of the Modern Synthesis – Mayr, Dobzhansky and Simpson – insisted that organic selection was at most of minor importance in evolution. In an influential article written in 1953, Simpson renamed organic selection 'the Baldwin effect' and, after giving a somewhat inaccurate version of the hypothesis, concluded that there was no good evidence and no theoretical need for it. Little more was heard about it until the 1980s, when interest in it and in Waddington's ideas was re-kindled as biologists began to use new mathematical, computational and experimental approaches to investigate how behaviour influences evolution.

1.6 Marginalized Empirical Data

The MS supporters' oft-repeated claim that empirical research has proved that soft inheritance does not occur had little experimental foundation even when the MS was constructed. The Institute for Experimental Biology (Biologische Versuchsanstalt, BVA) in Vienna, also known as the Vivarium, was founded in 1903 with the mission to '[e]stablish the role of external factors – temperature, lighting, an environmental medium, etc. – on changes in organic form, as well as on the persistence of those changes in descendants' (Przibram, 1903, p. 152).[6] In 1912, on the basis of nine years of extensive work, Hans Przibram, one of the founders of the BVA, was able to say that

> it was shown for all major groups of plants and animals that characters that were brought to appearance through environmental changes in the bodies of the parents can be maintained in the offspring raised under normal conditions. Now the question is open by which mechanisms the modification of the offspring was induced. (Przibram, 1912, p. 666).

By 1938, when the three Jewish founders of the BVA were deposed and Nazi scientists took over the institute (which was destroyed in 1945), 450 scientists from all over the world had visited or worked there, 358 different species of animals had been kept and 120 bred, and over 600 publications had been produced by the zoology department alone (Logan and Brauckmann, 2015). The investigation of developmental plasticity (phenotypic flexibility) was central to the research programme, which included studies of the altered development of ciliates and modified plant morphology following various treatments; of regeneration following injury; of the impact of chemicals on colour

---

[6] We thank Klaus Taschwer for the translation of this and the next quotation, and for other information about the BVA. See Müller (2017) for more details of the work and politics of the BVA.

adaptations in butterfly larvae; of the effects of high and low temperatures on the gonads, foetuses and next generations of mice and dormice; and of the impact of rearing temperatures on sexual characteristics and morphology in rats, which also included examination of transgenerational effects. The latter work was carried out by Paul Kammerer and the pioneering endocrinologist Eugen Steinach. Kammerer also studied the effect of background colour on the development and inheritance of salamanders' colouration, and the development of nuptial pads in the midwife toad.

Kammerer's work on the midwife toad is infamous. He made toads breed in water by increasing the temperature of their tanks, forcing them into the water to cool off. In just two generations, according to Kammerer, the male midwife toads, like their water-dwelling ancestors, had the black nuptial pads on their forelegs that gave them more traction during mating in water. Kammerer interpreted this as an acquired character brought about by adaptation to the environment. He thought that the effect was due to the activation of pre-existing genes, and claimed it was inherited. After World War I, during which the BVA was partially destroyed, the single remaining specimen was found to have been injected with black ink. Kammerer was accused of the fraud and committed suicide six weeks later, something that led to a general belief in his guilt, although his fraud was never proven. Klaus Taschwer believes that a far more plausible explanation is that the tampering was done by virulently anti-Semitic Austrian scientists.

Whatever the veracity of the midwife toad experiments, it is remarkable that the whole corpus of the BVA research has been almost totally forgotten. It seems as if Kammerer's suicide discredited all his results and tainted the entire research project of the BVA. This, together with the continued supremacy of former Nazi scientists in Austrian academies after World War II and the deliberate whitewashing of their Nazi past, conspired to bury all the BVA research.

In the post-war period, work on transgenerational inheritance of environmentally induced variation was also ignored because of its association with Lysenkoism. The tension between the West and the USSR led to what Karl Lindegren (1966) called the 'cold war in biology'. In the Soviet Union, for ideological reasons, Trofim Lysenko had been successful in suppressing 'bourgeois Mendelian-Morganist genetics' and having many of those who had been active in this formerly flourishing discipline dismissed (or worse). He published and encouraged fraudulent work on the inheritance of acquired characters, and replaced genetics with a distortion of science, which had devastating effects on Soviet agriculture. Lysenko was undoubtedly a charlatan and a tyrant, but, in spite of his dominance, there was a body of honest and careful research in the

USSR showing that the effects of environmental stress can be inherited and that Mendelian heredity cannot provide the full picture of biological inheritance. Studies such as those of Dmitry Belyaev and his group in Novosibirsk, which showed that 'dormant' genes may become heritably reactivated during the domestication of silver foxes, and Shaposhnikov, who showed heritable effects of changed nutrition in aphids, and many experiments showing the transgenerational effects of alcohol and morphine in animals, were little known in the West (for more details see Jablonka and Lamb, 1995). Most Soviet research was ignored in the USA, where the McCarthy witch-hunts of the early 1950s made biologists cautious about Soviet-promoted ideas for fear of being labelled communists.

In contrast, in Britain, where many of the leading geneticists were openly communists or sympathizers (although not Lysenkoists), unorthodox ideas about heredity were less of a problem. For example, at a Society for Experimental Biology Symposium held in 1952, papers were read on a wide range of subjects, many of which presented views and data that showed the inadequacy or incompleteness of the then-young Modern Synthesis. In later years, too, studies published by major scientific journals showed that environmentally induced traits can be inherited. For example, when the bacterium *Pseudomonas aeruginosa* was grown at 43°C for five generations it became much more susceptible to infection by bacteriophages, and this increased susceptibility persisted for 60–65 generations after it was returned to the normal temperature (37°C); the detrimental effects of temperature on pea growth were inherited for at least eight generations; changes induced by fertilizer treatment of flax plants were inherited for at least six generations; after growing on various fertilizers, changes in the flowering time and height of *Nicotiana* were transmitted for three generations; and experimental psychologists showed that many drugs and stressful treatments can have transgenerational effects that can sometimes last for several generations (for details, see Jablonka and Lamb, 1995, 2011; additional examples were documented by Lindegren, 1966). However, findings like these had very little effect on the MS version of evolutionary theory, which became increasingly dogmatic.

Something else that should have bothered those who were reinforcing the MS was the finding that sexually reproducing organisms do not always obey Mendel's rules. For example, Brink found that a maize allele affecting pigment formation can be semi-permanently modified by a second allele, a phenomenon he called 'paramutation'; McClintock's work on transposable elements showed that genes move around the genome, altering the developmental expression of neighbouring genes when they do so; and Crouse and many others recognized 'genomic imprinting', a type of inheritance in which the expression of a gene, a chromosome or a whole set of chromosomes depends on the sex of the parent from which it was inherited.

Work with microorganisms was also posing problems for conventional transmission genetics and altering ideas about inheritance. For instance, in some bacteria and protozoa, two alternative phenotypes of cells that are genetically identical and are kept in the same culture conditions can persist for many generations. Sewall Wright and Delbruck both suggested in the 1940s that steady-state systems are involved: alternative, self-regulating metabolic patterns are maintained because their components are transmitted to daughter cells during cell division. This was not Mendelian heredity! Similarly, inheritance in the protozoan *Paramecium* didn't fit Mendelian patterns. For instance, occasionally individuals form with two mouths, and this is inherited; when the system is manipulated so that a normal and a two-mouth individual swap nuclei and cytoplasm, the inheritance pattern is determined neither by the nucleus nor by the cytoplasm, but by the architecture of the cortex. Even differences caused by mechanical damage to the cortex are sometimes transmitted to daughter cells.

By the 1960s, it was also clear that in multicellular organisms development depends on soft inheritance. For example, dividing mammalian cells that are taken from different tissues retain some of their phenotypic features in culture; *Drosophila* imaginal disc cells (undifferentiated groups of larval cells that eventually form specific adult structures) that are serially transferred through adult female abdomens for many transfer generations usually retain their identity – their state of determination is inherited. Evidence that chromosomes can carry developmental information came from studies of X-inactivation: early in the development of female mammals, one of the two X chromosomes in each cell becomes condensed, late replicating and inactive. Once the decision is made, the same X remains inactive in all daughter cells.

Findings like these showed that cells with identical genes can produce different transmissible phenotypes. Yet this was not considered to be relevant to evolution, only to development. One mainstream evolutionary biologist who was aware of the potentially subversive implications of soft inheritance during cell differentiation was Maynard Smith, who wrote:

> The view generally taken by geneticists of differentiation, when it is not simply forgotten, is that the changes involved are too unstable to be dignified by the name 'genetic', or to be regarded as important in evolution. I tend to share this view, although I find it difficult to justify. (Maynard Smith, 1966, p. 71)

Some biologists did bite the bullet, however. In 1958, a few years after it had been accepted that the primary genetic material is DNA, David Nanney drew

attention to heredity based on 'epigenetic systems', which he and others thought
might underlie the differentiation and stability of somatic cell lineages. He used
the term 'epigenetic' to 'emphasize the reliance of these systems on the genetic
systems and to underscore their significance in developmental processes'
(Nanney, 1958, p. 712). His fellow microbiologist Boris Ephrussi accepted
Nanney's distinction between epigenetic mechanisms 'that regulate the expression of genetic potentialities' and truly genetic mechanisms 'that regulate the
maintenance of the structural information', but pointed out 'that, as a corollary,
we must admit that not everything that is inherited is genetic' (Ephrussi, 1958,
pp. 46 and 49). Nevertheless, the epigenetic inheritance systems of protozoa and
somatic cells, and ideas about their biochemical nature, were of scant interest to
evolutionists.

1.7 The Epigenetic Turn

Relatively little about epigenetics and epigenetic mechanisms is to be found in
the scientific literature of the 1960s and 1970s. The term was used adjectively
by biologists as a synonym of 'developmental', and continued to be employed
by microbiologists and by geneticists interested in variation. Waddington's
ideas, when not ignored, were explicitly condemned. In an influential book
published in 1966, George Williams, whose declared aim was 'to purge biology
of what I regard as unnecessary distractions that impede the progress of evolutionary theory' (p. 4), was damningly critical of Waddington's epigenetic
approach to evolutionary change. 'In explaining adaptation, one should assume
the adequacy of the simplest form of natural selection, that of alternative alleles
in Mendelian populations, unless the evidence clearly shows that this theory
does not suffice' (p. 5). Richard Dawkins' *The Selfish Gene* (1976) reinforced
the growing gene-centred view of evolution.

The work that eventually initiated the epigenetic challenge to this hard
version of the MS began with two theoretical papers published in 1975, one
by the British biologists Robin Holliday and John Pugh, and the other by the
American Arthur Riggs. Independently, these authors proposed a new type of
hereditary mechanism that could account for the transmission of cell phenotypes. They suggested that DNA modifications, specifically the methylation of
a site on cytosine (one of the four bases in DNA), could affect gene activity.
Methylation had no effect on the coding properties of the gene but affected the
binding of sequence-specific regulatory proteins. From work with bacteria, it
was known that patterns of DNA methylation could be enzymatically copied
during DNA replication, so daughter cells could inherit the same methylation
patterns and state of gene activity as the parent cells. In their highly speculative

papers, the scientists suggested that this could help to explain the switching on and off of genes during development, the stability of the differentiated state and its inheritance, heritable X-chromosome inactivation, genomic imprinting, transposition and various other developmental events. Also in 1975, Ruth Sager and Robert Kitchin also suggested that DNA methylation was associated with the selective silencing seen in imprinting and X-inactivation and with the regular elimination of chromosomes that occurred during the development of some insects. They, too, envisaged DNA methylation as an additional, heritable, information-transmitting system. In these pioneering papers, the evolutionary consequences of this type of information transmission were not explored.

Direct evidence for the inheritance of patterns of methylation in cell lineages was obtained in the early 1980s; usually, methylation was associated with gene silencing and non-methylated sequences with gene expression. As biochemical techniques for studying DNA methylation improved, details of the distribution of methylated cytosines on the chromosomes, the identity of some of the proteins that bind to them, and the changes in methylation patterns that occurred during development began to be worked out. As anticipated, it was found that there are widespread changes in methylation during mammalian gametogenesis and early development, when old epigenetic information must be erased and the genome reset for the new individual's development. Although not all the speculations in the 1975 papers received experimental backing, within a decade the basic idea that gene expression in some multicellular organisms is associated with patterns of DNA methylation was well supported experimentally.

In none of the seminal 1975 papers did the term 'epigenetics' appear, but in the next two decades 'epigenetic variation', 'epialleles' and 'epimutation', as well as 'epigenetic', were increasingly used in the context of cell heredity. One stimulus for this was Robin Holliday's 1987 'The inheritance of epigenetic defects', another wide-ranging and speculative paper in which he discussed how loss of methylation in somatic cells could lead to abnormalities in the newborn, to cancer and to ageing. He argued that most methylation faults in the germline are probably repaired, but he recognized that some epimutations would be transmitted to the next generation.

Two years later, we suggested that random and, more importantly, environmentally induced epimutations could have evolutionary effects, thereby adding a Lamarckian dimension to Darwinian evolution (Jablonka and Lamb, 1989). We argued that the mechanisms that lead to the transmission of cellular epigenetic variants – those discussed by Nanney as well as those discussed by Holliday – should be incorporated within an evolutionary framework that included the developmental reconstruction of *all* types of hereditary variations.

We called the processes, factors and mechanisms that underlie cellular inheritance 'epigenetic inheritance systems', defining an epigenetic inheritance system as 'a system that enables the phenotypic expression of the information in a cell or individual to be transmitted to the next generation' (Jablonka and Lamb, 1989, p. 292). In 1995 we reviewed the then existing cases of epigenetic inheritance and discussed the evolutionary implications of this type of heredity in *Epigenetic Inheritance and Evolution: The Lamarckian Dimension*. Our challenge to the MS was explicit, since our position was that soft inheritance is important in evolution.

Epigenetics was not the only field that was challenging ideas about inheritance and evolution at the close of the twentieth century. In the 1980s, theoretical studies of cultural inheritance and cultural changes in populations began to have an impact on thinking about human evolution, and in the 1990s the study of social learning in non-human animals began to take off. Consequently, in 2005 we put the ideas from these blossoming fields together in our book *Evolution in Four Dimensions*, in which we argued that heritable genetic, epigenetic, behavioural and symbolic variations are all important in evolution. Evolution, we suggested, should be defined as a change in the nature and frequency of heritable types (genotypes, epigenotypes and cultural types) over time.

2 Characterizing Inheritance Systems

In this Section, we outline the features of the different systems underlying the inheritance of phenotypic variations and give some experimental evidence for their existence. More details are given in Jablonka and Lamb (2014) and Lamm (2018). As a result of new data and fresh thinking, the classification we use here is slightly modified from that used in earlier publications.

2.1 Classifying Inheritance Systems

We define an 'inheritance system' as a set of evolved factors, processes and mechanisms that enables the developmental reconstruction of ancestral variations in descendants and typically results in similarity between them. Powell and Shea (2014) maintain that to qualify as an inheritance system, the processes involved need to have been *selected for* transmission. This criterion, they argue, disqualifies many types of between-generation transmission from being inheritance systems. We agree that the mechanisms behind genetic and most types of epigenetic, symbolic and some behavioural inheritance have additional functions. There are gradations between cases for which one feels confident that there has been strong selection for the transmission function (as with the genetic system), cases where transgenerational transmission is one of several functions (e.g. epigenetic systems) and cases where transmission may be an unselected by-product of other functions (e.g. some cases of behavioural inheritance). Even in the latter cases, however, we believe that there are sufficient similarities in the mechanisms and outcomes of the processes involved to warrant their description as inheritance systems.

Recognized inheritance systems include (i) the genetic inheritance system through which variations in nucleic acid sequences (usually DNA) are replicated; (ii) cellular epigenetic inheritance systems, which include transmission of variations based on self-sustaining feedback loops, three-dimensional templating, chromatin marking and RNA-mediated regulation; (iii) behavioural inheritance systems based on the transmission and reconstruction of non-symbolically mediated behaviours; and (iv) systems that are mediated by symbols, such as linguistic symbols. While the genetic and epigenetic inheritance systems are found in all living organisms, behavioural inheritance is confined to animals engaged in social learning, and the symbolic system is found only in humans.

All inheritance systems operate within developmental and ecological frameworks, and some animals leave legacies of their external environment or internal state as inputs for the development of their descendants or associates. For example, the ecological niches animals construct or modify are often inherited by their young. Similarly, elements of an animal's internal state, such as circulating hormones, food traces or their microbiomes, which vary from individual to individual, can be passed to descendants. In the case of the microbiome, the host is part of the ecological niche of its microbiota and the microbiota is part of the niche of the host, but from the perspective of the amalgamated unit (the 'holobiont') they form parts of its developmental system. The choice of perspective depends on the biological question asked and the degree of integration between the partners. In order to encompass these types of developmental-inheritable interactions, which include the 'infectious inheritance' of microbiomes, ecological legacies and the transmission of behaviour-inducing substances, we classify them as cases of 'soma-to-soma' or 'soma-mediated' inheritance. The latter term is probably better, since it can include not only gamete-independent transmission but also the transfer of elements transmitted from the soma into the gametes in every generation.

The variations generated and transmitted through different inheritance systems act as inputs into phenotypic traits at all levels of biological organization. The effects of different inputs interact, going in the same or opposite directions, and they can have additive or non-additive effects on the phenotype of interest. In Sections 2.2–2.6 we briefly describe each of the inheritance systems; their properties are summarized and compared in Section 2.7.

2.2 The Genetic Inheritance System

The genetic inheritance system is based on transmissible variations in nucleic acids, mainly in DNA. It includes all the factors and mechanisms that unwind, transport, replicate, repair, transpose, mutate, insert, delete, amplify, degrade and otherwise modify DNA sequences; on the chromosomal scale in sexually reproducing organisms it includes all the factors and mechanisms that segregate, pair, recombine and monitor chromosomes during meiosis; it also includes mechanisms brought into play following polyploidization, hybridization and other genomic upheavals. This fundamental and highly versatile system probably had its origins in an RNA-based inheritance system in ancient protocells, which was partially replaced in most organisms by the DNA system, although parallel RNA inheritance is still found in many taxa. It is generally accepted that all the processes of handling and replicating DNA evolved through selection for (among other things) its maintenance and transmission (Maynard Smith and Szathmáry, 1995). Until recently, DNA variation was thought to be the only basis for cumulative evolutionary change.

For a few years after Watson and Crick's discovery of the structure of DNA in 1953, it was believed that mutations are rare, unrepaired mistakes (i.e. accidents), most of which occur during DNA replication. This view was based on the observation that the fidelity with which the DNA sequences of genes coding for structural proteins are replicated is extremely high, and the occurrence of mutations seems to be unrelated to the developmental history of the individual or its ancestors. The discovery of several DNA repair systems reinforced the view that mutations are accidental changes that had escaped repair or been misrepaired. Some years later, however, it was discovered that mutation rates differ in different regions of the genome, that stress increases the rate of sequence changes (especially of transposition), and that recombination rates are not the same in all chromosome regions and can be influenced by stressful conditions. It was nevertheless assumed that the newly introduced DNA sequence variations are random with respect to function. The assumption that there is a stochastic element in the generation of DNA variations has been confirmed, but it has turned out that the extent of stochasticity can be under developmental control, and there are many evolved mechanisms that engineer non-random variations (see Shapiro, 2011, for many examples).

A dramatic example of an adaptive DNA engineering system, one which is changing the face of biological research, is the CRISPR (clustered regularly interspaced short palindromic repeats) system that protects prokaryotes against viruses and other foreign genetic elements. When alien DNA such as that from a virus is detected, a fragment of it is integrated into the CRISPR locus as a spacer. When the same virus re-infects the bacterium, the bacterium transcribes the CRISPR DNA, and the RNA sequences that are homologous to those in the DNA of the attacking virus bind to it and lead to its degradation. As Eugene Koonin has pointed out, this is a classical Lamarckian system: immunity to the invading virus is an adaptive acquired character that is inherited through the genetic system (Koonin, 2019).

2.3 Epigenetic Inheritance Systems

As with many other concepts, including gene, genotype and genetics, the meanings of 'epigenetics' and 'epigenetic' have shifted over time. Epigenetics is used in two main ways: in the wide sense, it refers to the study of interactions between genes and their products, which construct the phenotype. This is the original Waddingtonian sense of the term. In the narrower sense, one which is consistent with the usage of Nanney and other microbiologists in the 1950s and 1960s, epigenetics is 'the study of developmental processes in prokaryotes and eukaryotes that lead to persistent, self-maintaining changes in the states of organisms, the components of organisms, or lineages of organisms' (Jablonka and Lamb, 2014, p. 393). It is epigenetic mechanisms that provide the cellular memory of both dividing and non-dividing cells; they are what enable liver cells, kidney cells, neurons and other cell types to retain their functional identity. 'Epigenetic inheritance' is an aspect of epigenetics. The term refers to the transmission to subsequent generations of epigenetic variations that do not stem from differences in DNA base sequence. It has been found in all organisms in which it has been sought. Epigenetic inheritance systems are sets of cellular processes and factors that lead to inheritance of the state of activity of genes or the structure of proteins and other cellular components. Such inheritance can occur during somatic cell divisions and sometimes also following the sexual processes of gametogenesis. The mechanisms involved are diverse. Nevertheless, it is possible and useful to classify epigenetic inheritance systems according to their key properties. We have suggested that there are four major types of molecular mechanisms through which cellular epigenetic states can be passed to future generations:

1. *Steady-state systems* were the basis of David Nanney's recognition of transmissible epigenetic information. States of gene activity are perpetuated in lineages of cells or microorganisms because the components of feedback loops are transmitted to daughter cells. The feedback can occur at any stage of information processing. An example is the inheritance of two distinct alternative phenotypes, white and opaque, in the common human fungal pathogen *Candida albicans* (Hernday et al., 2013). In the white form, cells

are round and shiny, and form dome-shaped colonies on solid culture media, whereas cells of the less common opaque form are elongated and form dull, flatter colonies. The two forms also differ in the genes they express, the parts of the body they tend to colonize, and their interactions with their host's immune system. These differences are based on the activity of a master regulatory gene: when it is turned on, cells are opaque, and this state continues because the gene's product binds to its own regulatory region and maintains its activity; as cells divide, daughter cells inherit the gene product, thereby ensuring the positive feedback loop remains active and the lineage remains opaque. If, by chance or through an environmental change, the level of gene product falls, cells become white (the default state). The full picture is more complicated than this, involving additional genes and a network of interacting feedback loops that make the two alternative states more stable and persist for many (around 200) cell generations, but the principle is simple.

2. *Structural templating* is seen in the transmission of some cortical features of ciliated protozoa and in the building and transmission of complex biological membranes. It is best characterized in prions, which are misfolded proteins that tend to form clumps and multiply by a self-templating mechanism in which the misfolded molecules induce normally folded molecules to adopt their own variant form. During cell division, the prion particles are transmitted to daughter cells, where they continue to multiply (see Soto, 2012, for a brief review of protein inheritance).

   Prions are best known as the causative agent of transmissible neurodegenerative diseases such as bovine spongiform encephalopathy (BSE, or mad cow disease), Creutzfeldt–Jakob disease and kuru in humans, and scrapie in sheep. Humans or animals are infected by consuming prion-containing food or milk, by coming into contact with prion-containing urine or saliva, or by having blood transfusions or surgery that transfer prion particles. Prions occur in many other organisms too, including budding yeast and other fungi, bacteria, the mollusc *Aplysia*, the fruitfly *Drosophila* and the plant *Arabidopsis*. In many cases, the prion-associated phenotypes are potentially beneficial (see Section 3.2).

3. *Chromatin marking* involves DNA base modifications, such as methylation, or modifications of the histone proteins in chromatin that are associated with changes in gene activity. There is now overwhelming evidence that chromatin marks, which hitchhike on the DNA replication system, are sometimes transmitted through many generations of cells and organisms independently of any DNA variations. Some of the best examples come from studies of plants, in particular *Arabidopsis thaliana*, a small self-fertilizing weed. Using genetic

tricks, hundreds of inbred lines, all with normal and identical DNA sequences but different DNA methylation patterns, have been constructed. The variant epialleles (methylation variants) in these lines were shown to be inherited stably for at least eight generations. Moreover, some had phenotypic effects on traits such as flowering time, plant height and resistance to pathogenic bacteria, which in natural populations would affect their ability to survive and reproduce (Quadrana and Colot, 2016).

A few organisms, including the fruitfly *Drosophila*, the nematode worm *Caenorhabditis elegans* and yeasts such as *Saccharomyces cerevisiae*, which have all played major roles in genetic research, have genomes with little or no methylated cytosine. In these and many other organisms, inherited chromatin marks involve modifications of histones – proteins that are closely associated with the packaging of the DNA in chromosomes and hence in gene activity. For example, exposing *C. elegans* to a high temperature for five generations caused changes in gene expression that were associated with specific histone modifications and persisted for 14 generations (Klosin et al., 2017).

4. *RNA-mediated gene regulation* involves small non-coding RNA molecules that interfere with the processing of DNA information. They can do so in several different ways, but the end result is that specific genes are silenced or their activity is reduced. As well as having a role in gene regulation during development, RNA interference (RNAi) is important for silencing the activities of transposable elements and for neutralizing the effects of invading viruses and other types of foreign DNA.

The small RNAs responsible for gene silencing can be amplified and transmitted to daughter cells, and, in at least some plants and animals, they can also be transported to neighbouring cells, including germ cells (Dunoyer et al., 2013). Consequently, the changes in gene activity that they bring about can be transmitted to future generations. In *C. elegans*, for example, neuron-derived small RNAs mediate changes in the germline that result in the inheritance of two learned adaptive behaviours (pathogen avoidance and chemotactic responses) for up to four generations (reviewed in Charlesworth, Seroussi and Claycomb, 2019).

Small RNAs are abundant in mammalian sperm, and some have been linked to the inheritance of the effects of stress. Adult male mice that were traumatized early in life through temporary separations from their stressed mothers showed depressive-like behaviour and metabolic changes that were passed on to their offspring. Their sperm and some brain structures showed changes in the activities of small RNAs that are known to be involved in the stress responses and metabolic regulation. A causal link between these small RNAs and the behavioural and metabolic changes was demonstrated by injecting sperm RNAs from traumatized males into the eggs of normal mice: the resulting offspring showed behavioural, metabolic and molecular effects that were like those seen in traumatized males (Gapp et al., 2014).

What were the evolutionary origins of the four epigenetic inheritance systems we have described? While we accept that some of them may be mere by-products of selection for things other than transmission, in most cases we believe transmission has been one of their selected functions. As noted by Tikhodeyev (2018), almost any mechanism regulating gene expression or gene-product functioning may, under some circumstances, underlie heritable allelic differences. This is because alternative stable and heritable phenotypic variations (epialleles) can be formed whenever the regulatory system forms an autocatalytic feedback loop. We have argued previously that self-sustaining feedback loops were probably selected to ensure phenotypic continuity following cell division in unicellular organisms, so they were selected for between-generation transmissibility (Jablonka and Lamb, 2006). The fidelity of their transmission, like that of other heritable epigenetic states, would have been honed by natural selection to fit the temporal scale of fluctuations in the environment in which the organisms lived (Lachmann and Jablonka, 1996).

We suggested that the three-dimensional templating of molecular structures such as most membranes and large protein complexes was selected to allow rapid and reliable guided assembly rather than more error-prone and time-consuming self-assembly (Jablonka and Lamb, 2006). It can therefore be considered an evolved inheritance system. The same is true for chromatin marking, which probably originated through selection to ensure the stability of chromosomes and the continuity of gene expression patterns following cell division, as well as to silence genomic parasites.

RNA-mediated inheritance may have had its origin in the ancient RNA world, when RNA molecules replaced more limited and inefficient heredity systems, such as those based on metabolic networks, and became central in the metabolism, heredity and regulation of the first living organisms. In addition, RNA molecules may have had a role as a genomic immune system, defending cells against genomic parasites (see Jablonka and Lamb, 2014, pp. 326–7 for further discussion).

How do epigenetic variations at the cellular level map onto higher phenotypic scales? Variations in prions, self-sustaining loops, DNA methylation, histone modifications and small RNA profiles may have little or no effects at higher phenotypic levels, because the effects of many epigenetic variations are canalized. However, because they are involved in the regulation of gene expression and hence in the construction of cellular networks involving many genes, they can have, on the average, larger effects on the macroscopic phenotype than small localized DNA changes. Cellular epigenetic mechanisms are part of the regulatory response system, so variations induced by changed environmental conditions are likely to be less 'blind' to function than DNA variations, because they involve responsive regions of the genome. Some level of stochasticity is inevitable, however, and there seem to be cellular mechanisms that can harness this stochasticity. Soen and his colleagues (2015) suggest that intracellular selection among alternative biochemical networks can lead to 'adaptive improvisation' – a process of epigenetic exploration followed by developmental selection that can result in somatic and germline inheritance of the selected epigenetic variant.

2.4 Soma-Mediated Transmission

Gametic transmission is not necessary for the inheritance of phenotypes associated with epigenetic variations. As noted earlier, we use 'soma-to-soma' or 'soma-mediated' transmission as an umbrella term to describe the transmission of variations through the physiological reconstruction of the conditions in which organisms live and develop. Because of its scope and variability, it is difficult to provide a sharp characterization of soma-mediated transmission and to describe it as a distinct, evolved type of inheritance system, although each case may have been selected independently.

The epigenetic inheritance systems that we described at the molecular level always underlie soma-mediated transmission of phenotypes, but they are not sufficient for its full characterization. An example can illustrate this: female pups of rat mothers who gave them a lot of licking and grooming grow up to be adventurous and not readily stressed; they, in turn, give their own pups a lot of maternal care, so the cycle continues. Conversely, pups that are given less care grow up to be fearful and easily stressed, and show poor parenting, which leads to the perpetuation of the fearful behaviour in their own offspring. The outcomes of the differences in parental care are not due to inherited gene differences: they are the same even if a foster mother is used, so no information about parenting is transmitted through the germline. The information that leads to altered brain physiology in their young is transmitted through the mothers' behaviour, not through their gametes. At the molecular level, the differences between the maternal-care styles are correlated with differences in chromatin marks and gene activity in the hippocampus region of the brain, which are believed to be mediated by hormonal responses to the parental care given (Zhang and Meaney, 2010).

From this example, it is clear that understanding the transfer of the rats' maternal-care style requires a full description of the multi-level relations between the mother's care and her daughters' behaviour. For a behavioural variant, a complete explanation of its transmission has to include specification of the developmental (e.g. hormonal), ecological (e.g. niche constructing), microbiomic and social conditions that are reconstructed in each generation. In cases of microbiome transmission, the parents (often the mother) transfer the microbiome to the next generation, and its persistence is secured through both the reproduction of the microbes and the similarity of the selective conditions in the gut. This kind of transmission is not restricted to animals: in some plants, symbiotic fungi are transferred vertically from the mother plant to seeds, thus reconstructing the phenotypic variations (such as the toxicity of leaves) that result from the relation between the plant and the fungus (Hodgson et al., 2014).

2.5 Behavioural Inheritance through Social Learning

The behavioural inheritance system based on social learning is a special type of soma-mediated transmission. 'Learning' can be defined as the process through which inputs from the world or the body result in the storage of encoded information that changes the animal's disposition to respond to future exposures to the same or related inputs. 'Social learning' is learning that is enabled or facilitated by social interactions. It can lead to animal traditions and, as the nature and frequency of socially transmitted patterns or products of behaviour in a population change over time, to cultural evolution.

Social learning covers a broad range of mechanisms that include behavioural priming by transferred substances, such as hormones or food traces, as well as observational non-imitative and imitative learning (Avital and Jablonka, 2000). Like all types of learning, it is influenced both by evolved biases and by developmental biases that are the results of past learning. In most cases, especially in birds and mammals, behavioural transmission involves transmitting behaviour-eliciting factors that interact with observational social learning.

Non-imitative social learning increases the probability that an observing naïve individual does what the 'model' (usually an experienced individual) has done, although not how it did it. For example, young black rat pups in Jerusalem pine forests learn to strip pinecones and get at the seeds by being exposed to partially stripped cones left by their mothers. The pups do not copy their mother's behaviour, yet her presence and tolerance toward them when they snatch seeds or partially stripped cones from her are necessary for them to learn the stripping technique (reviewed in Avital and Jablonka, 2000). In this case, as in many others, the habitat provided by the parents ensures not only that the young are exposed to the same conditions as they themselves experienced, but also that their constructed niche facilitates niche-relevant learning by their young.

With imitative social learning, a naïve individual learns not only what to do (e.g. get food) but also how to do it. Among songbirds, parrots and cetaceans, vocal imitation is common and can lead to populations differing in their song dialects. Motor imitation in non-human animals was at one time thought to be much rarer than vocal imitation, but it has now been recognized in rats, chimpanzees, parrots and dolphins. Imitation is rarely totally blind and, like non-imitative learning, it is usually reconstructed and adjusted to the circumstances of the imitating animal.

Observational learning is based on cognitive mechanisms that are similar to those underlying individual (non-social) associative learning. Selection for increased attention to and tolerance of social others was likely to have been important during the evolution of social learning. Behavioural imprinting – the rapid and efficient selective learning from kin and neighbours, which is often restricted to short sensitive periods in early life – is probably the outcome of past selection for what to learn and remember, and when to learn it. It results in a persistent memory of who the animal's parents are, or what its habitat is like, or, in birds, what the local song dialect is, etc. (Avital and Jablonka, 2000; Laland, 2017).

Social learning can lead to animal traditions. We define an animal tradition or culture as a dynamic system or network of socially acquired, transmitted or reconstructed, stabilized and often selected patterns of behaviours, preferences, products of activity and modes of learning that characterize a community. Many species with cultures or traditions have now been identified, and the number and types of traditions recognized within a species have grown. For example, all cetacean species that have been sufficiently studied show culturally acquired behaviour for traits such as song, migrations, foraging behaviour, social conventions, cooperative associations with humans, and play; there is also evidence for culturally driven genetic evolution in this group (Whitehead, 2017; Whitehead and Rendell, 2014).

2.6 Symbol-Mediated Inheritance: the Symbolic Inheritance System

There is both continuity and a profound difference between human and animal cultural systems. In humans, it is symbolic representation and communication, and especially deliberate teaching, that construct cognition and are central to the mechanisms that underlie cultural learning.

We define a symbolic system as a communication and representation system in which the signs – the pieces of information transferred from sender to receiver or communicated to oneself – are components of a conventional, rule-bound system in which they relate to and refer to both objects and actions in the experienced world and other signs within the system (so that a sign is always part of a network of references). Full-blown symbolic systems are found only in humans, although some animals may have very domain-limited symbolic systems. Bee 'dance-language', for example, is confined to communication and representation only about food.

Human symbolic systems are formed through a process of social negotiation and cultural evolution, and provide a shared common ground on which a notion of an 'objective world' is formed (Tomasello, 2014). Prominent manifestations of symbolic systems such as literacy, moral law, art and science are products of complex modes of cognitive development and culture-driven reconstruction processes that involve teaching (Jablonka and Lamb, 2014; Laland, 2017; Sperber, 1996). The shared aspects of symbolic systems are most apparent and best studied in the linguistic system, which the linguist Daniel Dor (2015) suggested is a communication technology for the instruction of imagination. The cultural evolution of the linguistic system involved a long succession of gradual modifications, reversions and occasional technological revolutions, which altered hominin cognition and led to partial genetic assimilation of the human linguistic ability. How much and what type of genetic assimilation was necessary is an open question (Dor and Jablonka, 2010; Heyes, 2018).

Once in place, the linguistic symbolic system, which enables high-fidelity imitation and the transmission of information through deliberate teaching, allows very efficient transfer of information both within and between generations. It is the basis for the amazing cumulative cultures of humans.

2.7 Comparisons and Relations among Inheritance Systems

The inheritance systems described in this Section have many similarities and differences. These are best seen in Tables 1 and 2, which give a general (though inevitably simplified) overview. Table 1 describes the way in which information is reproduced; it shows (1) whether the organization of information is modular (= digital), so the units can be changed one by one (like nucleotides in DNA), or holistic (like a self-sustaining loop) with a change in one component destroying the whole, and whether heredity is based on replication or reconstruction; (2) whether or not there is a system or systems dedicated to copying or reconstructing that particular information; (3) whether or not information can remain latent (unused and unexpressed) but

Table 1 How information is transmitted.

Inheritance system or mode of transmission	Organization of information and type of hereditary process	Dedicated copying machinery[a]	Transmits latent information[b]	Fidelity of transmission	Direction of transmission[c]	Range of variation
Genetic	Modular; replicative heredity	Yes	Yes	Usually very high; under stress can be moderate	Mostly vertical	Unlimited
Epigenetic						
(i) Self-sustaining loops	Holistic; reconstructive heredity	No	No	Generally low; high when network is canalized	Mostly vertical	Limited at the single-loop level; unlimited at cell level?
(ii) Structural templating	Holistic; reconstructive heredity	Partially dedicated?	No	Variable; switching rate can be 10^{-5} for yeast prions	Mostly vertical	Probably limited at the level of the intracellular 3D complex
(iii) Chromatin marking	Modular or holistic; reconstructive and replicative heredity[d]	Sometimes (e.g. for DNA methylation)	Sometimes	Variable; can be 10^{-4} per CpG	Mostly vertical	Unlimited

Table 1 (cont.)

Inheritance system or mode of transmission	Organization of information and type of hereditary process	Dedicated copying machinery[a]	Transmits latent information[b]	Fidelity of transmission	Direction of transmission[c]	Range of variation
(iv) RNA-mediated	Modular; replicative heredity	Yes	No	Variable; three to four generations or greater	Mostly vertical	Unlimited?
Soma-mediated (organism-level legacies)	Holistic; reconstructive heredity	No	No	Variable	Vertical and horizontal	Limited for single legacy; broad for overall states
Behavioural						
(i) Transfer of priming factors	Holistic; reconstructive heredity	No	No	Variable	Mostly vertical but also horizontal	Limited (single behaviour level), unlimited (lifestyle)?
(ii) Non-imitative social learning	Holistic or modular;[e] reconstructive heredity	No	No	Variable	Vertical and horizontal	Limited (single behaviour level), unlimited (lifestyle)

(iii) Imitative social learning	Modular; replicative and reconstructive heredity	Possibly	Yes, several, including teaching	No	Variable	Vertical and horizontal	Limited for a single pattern, broad for overall repertoire
Symbolic	Modular and holistic; reconstructive and replicative heredity		Yes; through cultural and genetic inheritance		Variable; can be very high	Vertical and horizontal	Unlimited

Based on Jablonka and Lamb, 2014.

a The existence of a dedicated copying or reconstructing machinery is evidence for the evolved nature of the transmission mechanisms; when there is no dedicated copying machinery, the transmission function has to be studied for each case.

b Latency may come in degrees; for example, a self-sustaining metabolic loop can be self-maintaining, but if the concentrations of metabolites are low it may have negligible effects on the cell phenotype, whereas the same loop, under different conditions with a high concentration of metabolites, will have a distinct effect.

c There is far more horizontal transmission of DNA sequences than once thought, and infectious inheritance is quite common.

d For example, some chromatin marks (e.g. those involving DNA methylation) can be replicated base by base or reconstructed in blocks.

e Can be holistic when considered at the level of a single learned behaviour, but modular when sequential learning is considered.

? points to uncertainty due to limited information. For example, in yeast a single protein can assume many prion forms (Li et al, 2010).

Table 2 The generation and selection of heritable variation.

System or mode of inheritance	Selected for	Variation is blind or targeted/biased?	Developmental filters/selection?	Planning/ metacognition?	Involved in niche construction?	Enables open-ended cumulative evolution?
Genetic	Replication fidelity?	Mostly blind, some targeted	Selection among cells with genetic variations within an organism prior to reproduction	No (except for genetic engineering)	Only through its phenotypic effects	Yes
Epigenetic						
(i) Self-sustaining loops	Persistence of metabolic patterns following cell division	Mostly targeted	Selection among cells with different epigenetic states within an organism prior to reproduction; epigenetic states can be reversed or modified	No, for all epigenetic inheritance systems (although epigenetic engineering by humans may become an exception)	Yes, since the products of cellular activities can affect the environment in which the cell, its neighbours and its	Not as a rule, though it may occur
(ii) 3D templating	Persistence of complex structures following cell division	Mostly targeted but also blind				

(iii) Chromatin marking	Resistance to genomic parasites in parents and offspring; reconstruction of transcriptional states following chromosome replication	Mostly targeted but also blind	during gametogenesis and early embryogenesis			descendants live
(iv) RNA-mediated regulation	Resistance to genomic parasites; regulation	Mostly targeted but also blind				
Behavioural						
(i) Transfer of priming factors	Inducing disposition or behaviours	Mostly targeted	Yes, during development	No	Indirectly through their behavioural effects	Not as a rule
(ii) Non-imitative social learning	Social attention to experienced models	Mostly targeted	Yes, selection through learning	No, with the exception of some highly intelligent birds and mammals	Yes, social learning and traditions can alter social and ecological conditions	Not as a rule
(iii) Imitative social learning	Associating perceptual and motor representations	Mostly targeted, some blind	Yes, usually			

Table 2 (cont.)

System or mode of inheritance	Selected for	Variation is blind or targeted/ biased?	Developmental filters/selection?	Planning/ metacognition?	Involved in niche construction?	Enables open-ended cumulative evolution?
Symbolic	Shared normative common ground	Mostly targeted	Yes; cultural selection at many levels	Often, at many levels and in many ways	Yes, at many levels and in many ways	Yes

nevertheless be transmitted; (4) whether the fidelity of transmission is high (as is usual for the genetic system), variable or low; (5) whether information is passed only to offspring (vertically) or to neighbours as well (horizontally); and (6) whether variation at the functional-unit level is unlimited and capable of indefinite variation, or limited, in that only a few distinct variants can be transmitted.

Table 2 focuses on the generation and selection of hereditary variants during ontogeny and evolution. It shows (1) the type of selection shaping the inheritance systems; (2) whether new variants are blind ('random') or targeted to occur in specific structures, activities and functions; (3) whether they pass through developmental filters and are modified before transmission; (4) whether they are constructed by deliberate planning, which requires metacognition; (5) whether they change an environmental niche; and (6) whether they can lead to open-ended, cumulative evolution.

An important difference between the DNA inheritance system and most other systems of information transmission is that while the genetic system is based on replication, i.e. on dedicated mechanisms that copy a sequence irrespective of its function, most other inheritance systems are based on reconstruction, in which the state that is reproduced depends on the context in which it is reconstructed. For example, the reconstruction of the rats' mothering style described in Section 2.4 depends on the initial neurological and physiological conditions in the offspring, which were constructed by the extent of maternal licking and grooming they received. Often, both replication and reconstruction are involved in the regeneration of a variant. The reproduction of methylation patterns, for instance, is partially replication based: dedicated methyl-transferase enzymes copy patterns of DNA methylation, and which pattern is copied seems to be independent of function, yet, in many animals, methylation patterns are erased during gametogenesis or early embryogenesis and are reconstructed in the next generation. Similarly, with imitation, many patterns of behaviour, whatever their functional effects, can be imitated; usually, however, the process of imitation is not entirely blind, and patterns of behaviour that have functional significance for the individual are more readily imitated than others.

Together, what the two Tables tell us is that the dynamics of inheritance and of evolutionary changes that are based on the various inheritance systems are likely to differ. Models intended to describe these dynamics have to recognize this. They will depend on whether information is modular or holistic, whether variation is blind or targeted, how faithfully it is transmitted, whether it can remain latent, and whether and to what extent transmission is vertical or horizontal.

3 The Evolutionary Implications of Non-Genetic Inheritance

If there is more to heredity than DNA, then evolution can no longer be defined in terms of changes in gene frequencies. The DNA-variation-independent effects of non-genetic heritable variations have to be incorporated into evolutionary analyses. When selection is driving evolutionary change, we cannot assume that the process involves only the selection of genes, because selection of epigenetic and cultural variations, and their interactions with the genetic system, may also be involved. So, too, may be selection that is based on non-replicative processes of differential stabilization.

3.1 Arguments against the Importance of Epigenetic Inheritance in Evolution

Some passionate adherents of the MS, especially population geneticists, argue that the role of epigenetic inheritance in evolution is trivial at best.[7] The most common reasons given are: (i) epigenetic variation is just another source of heritable variation, which the MS can accommodate; (ii) almost all epigenetic changes occur in somatic cells, not the germline, and any that do occur in the germline are erased during gamete formation and early embryogenesis, when the genome is purged of all traces of the previous generations' developmental history; the few errors in erasure that persist have only minor evolutionary effects; (iii) the fidelity of epigenetic inheritance is too low for it to be the basis of evolutionary change; and (iv) there is no evidence that epigenetic inheritance affects adaptation or any other evolutionary processes in natural populations.

The first criticism, that epigenetic inheritance is just another source of variation and does not affect MS models, overlooks the importance of differences in the origin and transmissibility of variations. Epigenetic and other forms of non-genetic inheritance do not necessarily conform with Mendelian transmission rules or with the idea of the non-directed origin of new variants, both of which were core assumptions of the population genetics on which the MS was founded. Given the developmentally induced nature of some of the epigenetic changes that result in the inheritance of modifications, it must be acknowledged that there is a Lamarckian facet to Darwinian evolution. The range of new variation is, in part, functionally biased: unlike (mostly) random genetic mutations, new epigenetic variants often arise in response to environmental challenges, so they are functionally correlated with developmental conditions. For example, the epigenetic marks and factors associated with gene activity induced by salt stress are likely to be related to the evolved responses to salts, whereas

---

[7] See, for example, Charlesworth et al., 2017, and discussions of epigenetic inheritance in Jerry Coyne's blog 'Why Evolution is True'.

those induced by behavioural stresses are likely to be more related to the neurohormonal stress response, although some induced activities will be similar for the two types of stress. Moreover, when variation is induced by environmental conditions, many chromatin sites and epigenetic states can be altered semi-coordinately and a similar range of variants can be generated in many individuals at the same time. Consequently, induced heritable variants can spread through the population rapidly. This makes population-genetics models, which assume natural selection of genetic mutations that have low and functionally unbiased rates of variation, inappropriate for epigenetic evolution.

The second argument, that most epigenetic changes occur in developmentally segregated somatic cells and epigenetic marks are totally erased in the germline, has been shown to be wrong. First, in unicellular organisms, plants, and many animals, 'Weismann's barrier' is a fiction, because there is no early separation of the germline from the soma. Consequently, epigenetic variations can be transmitted when somatic cells become germline cells. In addition, memory factors such as small RNAs can be transmitted from soma to germline. This is known to occur in both plants and nematodes, and possibly in other taxa as well (see Section 2.3). Work carried out in the last two decades strongly suggests that the transgenerational transmission of epigenetic variations both through gametes and in ways that bypass gametes is not uncommon and may well be ubiquitous (Bonduriansky and Day, 2018; Jablonka and Raz, 2009).

The third argument, that the fidelity with which epigenetic variations are transmitted is too low to enable evolutionary change, is unsound for several reasons. The fidelity of transmission is certainly very variable, but equally certainly transmission can last for many generations, particularly in plants, where the extensive reprogramming of methylation marks that is characteristic of mammalian gametogenesis and embryogenesis does not occur (Quadrana and Colot, 2016). For example, in *Arabidopsis thaliana*, where epigenetic inheritance can involve tens of thousands of differentially methylated CG sites in the genome and thousands of differentially methylated regions, the epimutation rate is $\sim 10^{-4}$ per CG per generation. This is low enough to enable significant cumulative evolution (Van der Graaf et al., 2015). Additional evidence for the stability of some epigenetic variants comes from experiments with the fruitfly *Drosophila*, where induced alternative epialleles that differ in specific histone modifications have been transmitted in genetically isogenic lines for over 50 generations. These epialleles can induce paramutation, can show parent-of-origin effects, can be reset to their naïve state by disruption of long-range chromatin interactions, and their expressivity can be modulated by environmental temperature changes (Ciabrelli et al., 2017). Some of the best evidence for stable epigenetic inheritance comes from yeast (*Saccharomyces cerevisiae*), where self-templating prion-like proteins, which have been shown to have adaptive value, can persist for over 100 generations (Chakrabortee et al., 2016).

A further reason why there is need for caution when thinking about transmission fidelity is that when epigenetic variation is developmentally selected as well as developmentally induced, the rate of adaptive evolution can be enhanced even if the fidelity of transmission is relatively low. An intuitively clear analogy is with cultural technological innovations: because new models are continually invented, tested and selected in research and development units before they are released onto the market, the persistence of any particular marketed model may be low simply because it is replaced by a new model already tested and selected in the R&D unit. Moreover, even with relatively low transmission fidelity, population-epigenetic models show that inheriting induced variants leads to very different population dynamics from those predicted when genetic inheritance alone is considered (reviewed in Bonduriansky and Day, 2018).

How faithfully induced epigenetic states are transmitted to future generations clearly depends on the organism, the mechanism of transmission, the phenotypic traits affected and the environmental conditions. This is to be expected if fidelity has been shaped by natural selection: inheriting a parent's phenotype has survival value for a lineage if the conditions that the parent encountered and adapted to persist, because the offspring are pre-prepared; but if conditions become different from those of the parent (which is likely in long-lived species), inheriting their adaptations would be detrimental. Natural selection acting on the dynamics of the establishment, maintenance and erasure of epigenetic states will therefore adjust the fidelity with which they are transmitted according to the probable persistence of the selecting conditions. For example, selection is likely to increase the stability of the inheritance of chromatin marks that suppress the activities of transposons, whose movements can have major and often disruptive effects on the genome and hence the phenotype; conversely, reducing the fidelity of transmission will be favoured if epigenetic changes result in adaptations to temporary conditions, such as those presented by rapidly evolving parasites. A more general reason why the fidelity of transmission will be adjusted through selection is one that applies to all taxa and may explain why epigenetic inheritance seems to be ubiquitous: it is the advantage of epigenetic priming – the inheritance of altered *dispositions* to respond adaptively to stimuli experienced by ancestors. If an induced change in gene expression leaves a mark that alters the threshold of future responses to that inducer, epigenetic inheritance of this primed state is likely to be beneficial: it does not lead to inappropriate responses, because the environmental stimulus is still necessary, but the speed and efficiency of response are heritably altered. The advantage of rendering the expression of a gene more or less likely following earlier induction in an organism's parents or more distant ancestors is similar to the advantages of neural learning, especially sensitization and habituation.

The fourth and most frequent reason given when MS adherents claim that epigenetic inheritance does not play a significant role in evolution is that there is no evidence for it. In Sections 3.2–3.6 we look at some of the effects of epigenetic inheritance that challenge this claim (see also Jablonka, 2017).

3.2 Adaptive Evolution Mediated by Epigenetic Inheritance

Mathematical and simulation models show that evolutionary dynamics in populations are altered when epigenetic variations are considered. To account for the changes in the frequency of epialleles in a population, factors such as the probability that environmental conditions will induce an epimutation, the paramutation rate (likelihood that the epigenetic state of one allele alters the state of another allele), and reset coefficient (the probability of erasure of the epigenetic state shortly before or after transmission) must be added to parameters such as selection, drift, gene flow and the stochastic epimutation rate. Considering all these parameters and their interactions leads to rates and distributions of phenotypic change that are very different from those expected on the assumption that inheritance is based only on genes (reviewed by Bonduriansky and Day, 2018; Jablonka and Lamb, 2014). One important prediction is that adaptations to new conditions are likely to start with heritable epigenetic variations rather than genetic variations (Pál, 1998). Another is that epigenetic variations can be beneficial in fluctuating conditions that last longer than the generation time of the individual but not long enough to enable genetic fixation (Lachmann and Jablonka, 1996). Rivoire and Leibler (2014) show more generally that, contrary to MS-based intuitions, theoretical models indicate that soft inheritance can be adaptive in many conditions.

Direct empirical support for epigenetic-based, adaptive evolutionary change in natural populations is not easy to obtain because of the confounding effects of genetic variation. For this reason, clonal and parthenogenetic organisms are often used when the effects of epigenetic variations are investigated. However, there is evidence from natural populations of several sexually reproducing plant species ranging from mangrove trees to Mediterranean violets, and from wild animals including fish, bats and sparrows, that the epigenetic diversity in methylation marks in different populations is greater than the genetic diversity in comparable regions of DNA, and some studies show a correlation between the variations in marks and environmental factors (reviewed in Hu and Barrett,

2017; Richards et al., 2017). For example, some of the DNA methylation differences among clonal Japanese knotweed plants that were collected from different habitats and then grown in a common environment were correlated with their original habitat, suggesting that environment-induced methylation changes had persisted through clonal propagation. Many animal species also show stable population-, habitat- or species-specific DNA methylation patterns, but the causal role of these variations is difficult to establish.

More direct evidence for the involvement of heritable epigenetic variations in traits that are relevant for fitness has come from experimental studies of isogenic *Arabidopsis* lines with hundreds of differentially methylated regions in their genome. Several of these regions were found to act as epigenetic quantitative-trait loci that accounted for 60–90 per cent of the heritability for two complex traits: flowering time and primary root length. A significant fraction of the differentially methylated regions that were studied also vary in natural populations, suggesting that such epigenetic variants could provide the basis for Darwinian evolution (Cortijo *et al.,* 2014).

Heritable, stress-induced epigenetic variation has been found to be involved in long-term adaptation in rice. When two varieties in which genetic variation was almost completely absent were exposed to drought for 11 generations, the drought tolerance they evolved was associated with the non-random appearance of heritable changes in DNA methylation, which accumulated over time. Many of the epialleles were in genomic regions that participate in drought-responsive pathways (Zheng et al., 2017).

How epigenetic variation affects adaptation over a longer timescale (about 200 generations) has been studied using asexual lines of the unicellular green alga *Chlamydomonas reinhardtii*. These were grown in four different conditions (control, salt stress, phosphate starvation and CO_2 enriched). When the amount of epigenetic variation (DNA methylation) was reduced through genetic or chemical manipulations, adaptation to environments where it would otherwise have happened was diminished. The changes in methylation affected some categories of genes more than others but did not appear to be linked to genetic mutations, suggesting that in this species and on this timescale, transgenerational epigenetic effects play a role in adaptive evolution (Kronholm et al., 2017).

Some of the clearest evidence for the adaptive role of heritable epigenetic variants comes from studies of prions and prion-like proteins. These polypeptides can adopt two or more conformational states, at least one of which reproduces within cells by self-templating. They are usually transmitted to all daughter cells, including meiotic progeny. Many proteins with different structures and functions can form prions, and they have been found in organisms ranging from bacteria to humans. The ability of a particular protein to adopt alternative conformations, which are often associated with different pheno-types, seems to have been conserved in evolution, even when the protein's amino acid sequence diverges. This suggests they have an adaptive function. Studies of protein-based inheritance in yeast have provided substantial evidence for this. Spontaneous switching between the normal and prion conformations (the equivalent of 'mutation') in yeast occurs at low frequency (10^{-4}–10^{-7}), but switching is greatly enhanced by stresses such as heat, osmotic shock, or high or low pH, all of which disrupt protein homeostasis. Chakrabortee and her collea-gues (2016) induced about 50 heritable prion-like conformations in yeast by transiently overproducing their proteins. The induced prions persisted for hundreds of generations after expression levels returned to normal, and many had beneficial effects when cells were subjected to stresses.

Though still small, the number of empirical studies showing that epigenetic variations are involved in adaptive evolutionary change is growing. When induced heritable variation is coupled with both developmental (intra-organism) selection and classical natural selection in populations, it can lead to rapid, developmentally guided adaptations. Braun and his colleagues have provided an example of what can happen by following the events that occur during the adaptation of bakers' yeast, *Saccharomyces cerevisiae*, to a totally new stress (reviewed in Braun and David, 2011). They used a genetically engineered haploid strain in which an essential gene that is needed to produce histidine was given the promoter of a gene from the galactose utilization system, which is strongly repressed in glucose medium. When such rewired cells are placed in a glucose medium without histidine, they cannot produce histidine. For them to grow and reproduce, a novel adaptation is required. What Braun and his colleagues found was that, after a lag period of 6–20 days, about half the cells started to multiply. In these cells, the regulation of the promoter was altered, and this change was inherited for hundreds of generations. In most cases, the basis of the altered regulation system was not mutation; it seems to have involved complex global rewiring of metabolic circuits, with different cells finding different adaptive solutions. The plasticity of cellular gene regula-tion allowed many adaptive solutions to be reached.

Induction followed by intracellular and intra-organism developmental selec-tion, which is accompanied by natural selection among the developing indivi-duals, is probably not uncommon. It may have occurred during animal domestication. Evidence for this comes from the rapid evolution observed in an experiment to domesticate silver foxes, which was initiated by Dmitry Belyaev in Novosibirsk in 1958 during the Lysenko period. Because silver-fox fur was a valuable export product for the USSR, and cage-bred wild foxes were difficult to handle, Belyaev was able to convince the authorities that trying to domesticate silver foxes was a worthwhile project. It was the beginning of a brave, successful, long-term evolutionary study that has now been running for 60 years. As well as making the foxes very human friendly, selecting solely for tameness rapidly led to faster development; there were changes in the foxes' reproductive pattern, in pigmentation, in the shape of their tails, ears, snout and legs; their vocalizations altered; B chromosomes (supernumerary, non-essential chromosomes) became more frequent; the pattern of inheritance of a pigmentation pattern was found to be non-Mendelian; and the levels of corticosteroids and neurochemicals such as serotonin were altered (reviewed in Jablonka and Lamb, 1995; Markel and Trut, 2011). Using a framework of thinking that had been developed by Ivan Schmalhausen (see Section 1.5), Belyaev reasoned that the stress imposed by selection upset the whole neuroen-docrine control of ontogenesis: previously integrated genetic systems were destabilized and 'dormant' genes were activated, producing multiple phenotypic effects. Today, it is easy to relate these ideas to what we know about the hormonal regulation of epigenetic changes.

Heritable epigenetic modifications were probably also involved in the domestication of the chicken, which began about 8,000 years ago. When compared with their red junglefowl ancestors, chickens show massive, genome-wide, heritable changes in methylation in the brain and other tissues. The extent to which these are dependent on DNA sequence differences is not clear (Nätt et al., 2012). However, selection for fearful and non-fearful behaviour in the junglefowl for only five generations led to divergent DNA methylation in 22 genomic regions in the hypothalamus, some of which were associated with neural functions and cellular metabolic pathways relevant to the stress response (Bélteky et al., 2018).

3.3 Interactions between Epigenetic and Genetic Inheritance, and the Phenotype-First Perspective

Evolutionary change can involve both genetic and epigenetic inheritance, although their relative importance depends on the timescale being considered. The context-sensitive and more frequent epigenetic variations may often initiate evolutionary change and then bias and facilitate long-term genetic changes, but throughout, epigenetic and genetic inheritance are interacting. Both direct and indirect interactions are involved, and both are important in evolution (Jablonka and Lamb, 1995, 2014). Direct interactions occur when changes in the structural elements of one inheritance system affect another system; indirect interactions are mediated by selection.

Some of the direct effects of genetic changes on heritable epigenetic variations are obvious: for example, if the number of CG sites in a stretch of DNA increases or decreases, it changes the probability that this region will carry methylation marks. The effects of some other genetic changes are less obvious: hybridization between species often leads to systemic alterations in methylation and RNA silencing; DNA deletions and insertions can lead to epigenetic silencing of allelic sites during meiosis; and mutations in genes affecting the epigenetic machinery (e.g. the DNA methylation or RNAi machinery) can alter patterns of epigenetic variation and the fidelity of their inheritance. The converse is also true: epigenetic modifications can have direct genetic consequences. If the cytosine of a CG site in DNA is methylated, it mutates to thymine at a far higher rate than that found with other transitional changes; epigenetic modifications of transposable elements or their regulators change the likelihood that they will move to other sites in the genome; and epigenetic marks affect recombination rates. All these direct interactions can influence the rate of genetic evolution.

The indirect effects of epigenetic variations on the rate and direction of genetic evolution are far-reaching because they alter the exposure of previously cryptic genetic variations to selection. They can lead to genetic assimilation, the process studied theoretically and experimentally by Conrad Waddington in the mid-twentieth century (see Section 1.5). Genetic assimilation occurs when, through selection, a facultative feature (one induced by environmental conditions, like skin calluses induced by rubbing) becomes constitutive (independent of environmental induction, like the calluses on the rump and sternum that ostriches have at hatching). Mary Jane West-Eberhard (2003) suggested a more general term, 'genetic accommodation', which encompasses not just the evolution of more canalized responses but also the evolution of greater plasticity – of increased dependence on environmental stimuli. As West-Eberhard sees it, evolutionary change starts with phenotypic accommodation: when an organism encounters an environmental or genomic challenge (e.g. a new mutation), it copes using existing mechanisms and properties of the developmental system, such as mechanical flexibility, turning gene networks on and off, and selectively stabilizing beneficial outcomes of cellular, physiological and behavioural exploration. These processes lead to correlated changes in various aspects of the phenotype. Natural selection for the best responders – for the best phenotypic accommodators – follows and leads to genetic accommodation, in which the basis for the selected differences between organisms is genetic. According to this scenario, whatever the nature of the challenge organisms face, the developmental-epigenetic response comes first, and genetic changes stabilizing or fine-tuning the developmental change follow. This is why West-Eberhard insists that genes are followers in evolution, not leaders.

Neither Waddington nor West-Eberhard considered the evolutionary effects of the inheritance of epigenetic changes. Yet, because we know that changes induced in inbred lines can be inherited (Section 2.3), not all cases of genetic assimilation and accommodation can be attributed to the presence of cryptic genetic variation. Where inheritance of an accommodated state depends on epigenetic variation, the process of adaptation could be far more rapid. Stajic and her colleagues (2019) have shown how the rate of genetic evolution in yeast is affected by inheriting epigenetic silencing. They constructed strains that were genetically identical apart from the chromosomal location of a gene (*URA3*) essential for the synthesis of uracil. When this gene is ON (active), cells can synthesize uracil, but they die in the presence of the drug 5-fluorootic acid (5-FAO); when it is OFF (silent), the cells require uracil but are resistant to the drug. Because the chromatin structure in different chromosome regions varies, the location of the gene heritably modifies the frequency with which cells switch stochastically between ON and OFF states and therefore the average degree of silencing in the population. By challenging populations having different degrees of silencing to grow in medium containing the drug 5-FAO, it was found that those with no epigenetic silencing adapted very slowly – it took a long time for new life-saving mutations to accumulate. Populations with a high degree of silencing adapted to the drug rapidly, because the gene switched to OFF frequently, and cells with this epigenetic silent state had a selective advantage. Genetic changes that silenced the gene were not apparent until much later, because competition with the epigenetically adapted cells impeded their fixation. Populations of cells with a low or intermediate level of silencing increased in size as more and more cells switched to OFF and these heritably silenced cells outcompeted others. As a result of the increased population size, there were more targets for mutation, and genetic assimilation of the silent phenotype was rapid. Some mutations were in the *URA3* gene itself, but others were thought to be in the silencing machinery that made the inheritance of the OFF epigenetic state more robust.

These experiments with yeast showed convincingly that selection of a heritable chromatin-based epigenetic variation can lead to genetic assimilation by increasing the population size and hence the number of mutations. A process we called 'mutational assimilation' leads to similar effects (Jablonka and Lamb, 1995, pp. 167–71; Pocheville and Danchin, 2017). Other experiments with yeast have shown how epigenetic inheritance through structural templating can lead to genetic assimilation. The best-known example involves the prion form of the yeast translation termination factor Sup35. In its normal form, this protein helps recognize the stop codons in mRNA that signal the end of the polypeptide. When the protein is in its prion form, the stop codons are ignored, thereby allowing the DNA sequences beyond them, which are usually not translated, to be expressed and have phenotypic effects. Because many genes have non-translated sequences that are normally cryptic, prion-containing cells can produce a burst of new phenotypic variants as the formerly silent regions are expressed. Even if prions are present in very few cells in a population, because they are self-perpetuating and inherited, the variety of phenotypes they produce can enable populations to survive and grow in adverse conditions. If these conditions persist long enough, a new beneficial epigenetic trait could be genetically assimilated through, for example, selection of genetic variants in the prion-forming proteins which mimic the epigenetic effects.

Soen (2014) has suggested another type of assimilation, one that recognizes the reciprocal dependence between an animal and its microbiome. He argues that when a stressful environment disrupts the normal epigenetic, metabolic and physiological functions of an animal, thereby destabilizing its phenotype, the microbiome with which it has coevolved is also affected. However, the size of the microbial population, the variety of species it contains and their short generation times mean that the microbiome adapts to the stress faster than the host. As the old coadaptation between host and microbiome breaks down, it further destabilizes the host. In this way, the microbiome promotes host phenotypic variability. Because the microbiome is inherited, host phenotypes can persist without genetic change until a new coordinated state of host phenotype and microbiome is reached. This phenotype may then be genetically assimilated.

3.4 Epigenetic Inheritance and Macroevolution

The effects of non-genetic inheritance systems on genetic evolution go beyond microevolution: they also play an important role in macroevolution. In Section 1.2 we described how, in the early days of the MS, Goldschmidt, one of the MS dissidents, argued that the emergence of new species and higher categories could not be explained by slow, cumulative microevolution. His claim was that they needed 'systemic mutations', something that Huxley (1942, p. 457) regarded as 'unproven and unnecessary'. Today, however, it is quite widely acknowledged that the rapid speciation Goldschmidt wanted to account for is far more common than once thought. In particular, it is now recognized that the genomic stresses resulting from hybridization can initiate a rapid, systemic, genetic and epigenetic reorganization. This can trigger epigenetic mechanisms resulting in meiotic silencing, alter the epigenetic regulation of multiple loci, and create new gene regulatory circuits that lead to new phenotypes (reviewed in Jablonka and Lamb, 2014).

Traditionally, speciation is seen as the result of genetic divergence during periods of geographical or ecological isolation, but, especially when exposed to new conditions, populations will accumulate random and induced epigenetic variations far more rapidly than genetic variations. These epigenetic changes may result in some degree of reproductive isolation. For example, changes in genomic imprinting or its loss can lead to reproductive isolation through hybrid inviability, because paternally and maternally imprinted genes are incompatible. There is evidence of this in the rodent genus *Peromyscus*, and on the basis of his studies of hybridization in this and other mammals, Vrana (2007) suggested that interactions among imprinted genes have played a significant role in generating mammalian species diversity.

Genomic imprints are not the only type of epigenetic variation that is significant in species divergence. Smith and her colleagues (2016) investigated populations of darter fish that may be in the initial stages of speciation. They found that epigenetic divergence between the populations was far greater than genetic divergence. Moreover, when closely related species were compared, epigenetic divergence was a better predictor of behavioural isolation (the strongest reproductive barrier in the genus) than genetic divergence. A similar conclusion was reached when comparing genetic and epigenetic divergence in different species of Darwin's finches in the Galapagos Islands (Skinner et al., 2014).

It is possible that epigenetic variations played a role in the evolution of our own species. The new field of paleo-epigenetics, which is based on methods enabling DNA methylation in ancient genomes to be reconstructed, has allowed comparisons between the methylomes of modern and archaic humans (Neanderthals and Denisovans) and the identification of epigenetic variations that may have influenced human evolution (Gokhman, Meshorer and Carmel, 2016).

In natural populations, it is difficult to unravel the contributions that genetic and epigenetic divergence make to reproductive isolation, but there is evidence from laboratory work with *Arabidopsis* that epigenetic differences alone can create widespread genomic changes in hybrids. Crossing two genetically identical but epigenetically very different lines, one with greatly reduced DNA methylation and one normally methylated, resulted in 'epihybrids' with many new methylation patterns that were not present in the parents and massive changes in gene expression; many transposons had decreased methylation, which could result in transposon movements in the progeny (Rigal et al., 2016). Hence, even without genetic change, in hybrids between two populations that have been geographically or ecologically isolated, epigenetic incompatibility between the parental genomes could bring about phenotypic change and initiate reproductive isolation between the parental populations.

3.5 Epigenetic Inheritance and the Maintenance of Genome Integrity

It has been argued that both the methylation marking and RNAi systems evolved primarily to defend cells against foreign or rogue DNA sequences. Although we do not entirely agree with this, because we believe that right from the beginning these epigenetic systems also had a role in cellular regulation (Jablonka and Lamb, 2014, pp. 323–7), it is certainly true that present-day organisms use epigenetic systems to generate adaptive, heritable responses to genetic elements that threaten genome integrity. For example, Rechavi (2014) has described how in the nematode *C. elegans*, the fruit fly *Drosophila* and other organisms, heritable small RNAs generated in response to attacking viruses or transposons serve as 'inherited vaccines' that protect against future invasions. The small RNAs are amplified, can move between different cells and tissues, including from somatic cells to germline cells, and can be inherited for hundreds of generations.

The ways in which epigenetic mechanisms are employed in cellular defence are not the same in all organisms, and even within a species several different processes may be present. In the filamentous fungus *Neurospora crassa*, for example, there are at least three ways in which transposable elements or introduced DNA can be silenced (see Aramayo and Selker, 2013, for details). 'Repeat-induced point mutation' (RIP) occurs in mitotic cells prior to sexual reproduction; it involves extensive cytosine-to-thymine mutation and concomitant methylation of most other cytosines in duplicated regions. 'Quelling', which suppresses the spread of transposons during vegetative development, involves small RNAs. It was discovered when in around 30 per cent of the transformants produced by introducing foreign DNA into the fungus, not only was the transforming DNA silenced but so too were native sequences homologous to it. Another process, 'meiotic silencing', operates only in meiosis and results in chromosome sequences that are unpaired being heritably silenced. It involves small RNAs and protects against the effects of errant transposons.

Many of the details of these epigenetic processes remain to be worked out, but an awareness of their existence is important for evolutionary biologists because it opens up new ways of looking at some evolutionary problems. For example, the silencing of unpaired chromosome regions during meiosis could have played an important role in the evolution of heteromorphic (morphologically different) sex chromosomes (Jablonka, 2004a; Jablonka and Lamb, 1990, 1995).

Epigenetic modifications and epigenetic inheritance are important not only in cellular defence but also in parasite–host interactions. Both parasites and hosts need flexible systems that can respond rapidly and reversibly to the challenges they pose for each other. Lachmann and Jablonka (1996) suggested that parasites use epigenetic mechanisms to evade their host's immune systems, and hosts used epigenetic mechanisms to transfer successful defences to offspring or prime their response to the pathogens. There are studies that support these predictions. Vilcinskas (2016) has shown how epigenetic mechanisms are involved in acquired resistance and priming of offspring and grandoffspring of the red flour beetle and the greater wax moth to fungal and bacterial pathogens. DNA methylation, histone modification and small RNAs were involved in the transmission of the resistance.

3.6 Epigenetic Inheritance Systems and the Evolution of Ontogeny

Epigenetic inheritance, which has been found in every unicellular organism in which it has been sought, was crucial for the evolutionary transition to complex multicellular animals, fungi and plants. During development, cellular states (e.g. whether a cell is a blood-forming cell or a liver cell) change as cells become determined and differentiated, and these states have to be remembered and passed to descendant cells. Because epigenetic systems are involved in both cellular control and cell memory, they confer plasticity on the developing individual: cells can respond to changing conditions, and, through somatic selection among different variants, ontogenetic adaptation is possible.

Epigenetic systems therefore have to be flexible, but they cannot be sloppy, because switching to alternative states could disrupt the division of labour within the organism and lead to it functioning less efficiently. This is what occurs in cancer. So once a lineage's fate is determined, selection will favour making epigenetic inheritance as reliable as possible. However, although stability is important, the production of the next generation of individuals would be impossible unless some cells retain or have the capacity to adopt an uncommitted state. We have argued that many aspects of development can be interpreted as outcomes of selection against carrying irrelevant epigenetic information into the next generation (Jablonka and Lamb, 1995, 2014). The most obvious is the early segregation of primordial germ cells, which is a feature of some animals' development: if germline cells are separated physically and divide rarely, there will be few epigenetic memories to carry into the next generation. In addition, most epigenetic memories can be eliminated during the genomic reprogramming that occurs during meiosis and early embryogenesis. This focus on epigenetics led us to conclude:

The efficiency of cell memory, the stability of the differentiated state, selection and cell death among somatic cells, the segregation between somatic and germ-line cells in some animal groups, and the massive restructuring of the chromatin of germ cells are all partly shaped by the selective effects of EISs [epigenetic inheritance systems]. (Jablonka and Lamb, 2014, pp. 249–50).

The inherited epigenetic marks that result in genomic imprinting, where the expression of a gene depends on the sex of the parent from which it was inherited, has also had evolutionary effects on development. These marks probably originated as by-products of the selection that led to the different ways DNA is packaged in sperm and eggs (Jablonka and Lamb, 1995). The resulting differential expression of maternally and paternally derived alleles in the embryo was then eliminated, diminished or enhanced by further selection that affected chromatin structure as imprints were recruited and modified for different developmental functions. In several unrelated groups of insects (e.g. some dipterans and coccids), imprints are the basis of sex determination; in mammals, they are significant in dosage compensation; they are thought to have had a significant role in the origin of flowering plants; and in both eutherian mammals and flowering plants, they have roles in the processes that allocate resources to developing offspring (Jablonka and Lamb, 2014; Rodrigues and Zilberman, 2019).

Epigenetic memory and inheritance mechanisms were also important for the evolution of learning, a central developmental strategy of neural animals. All the four epigenetic systems we discussed in Section 2 are used in neurons and are fundamental to animal learning (Bronfman et al., 2014). When open-ended associative learning first evolved in neural animals, it probably led to over-learning, and overlearning led to chronic stress, a deleterious effect that was exacerbated by the transmission of the effects of stress to offspring. Ginsburg and Jablonka (2019) have suggested that the evolution of the control of memory and of forgetting was driven by the effects of stress-associated learning.

3.7 The Evolutionary Effects of Social Learning

Twentieth-century studies of cultural transmission and of cultural change as a Darwinian evolutionary process were initiated by Cavalli-Sforza and Feldman (1981) and Boyd and Richerson (1985). They constructed models that followed the fate of variant cultural traits (patterns of behaviour, beliefs, attitudes, skills, etc.) that are transmitted among individuals though learning. Usually their models described human cultural evolution, but they can equally well be applied to cultural change in non-human animals. Like population genetic models, they explore the effects of population size (cultural drift), migration (of people or their products), mutation-like processes (accidental innovations) and selection on the frequency of transmitted cultural variants in populations. In addition, the models include the possibility that cultural variations can blend, and that transmission is not only (or mainly) vertical but also horizontal (among peers) and oblique (to unrelated members of the next generation). Factors that bias learning, such as learning from prestigious and/or successful individuals, conforming by learning from the majority, preferring to learn certain things because of cognitive biases or instilled habits, and the effects of individual trial-and-error learning, all of which influence which cultural patterns are acquired and transmitted and the fidelity with which they are transmitted, are also incorporated in the models. These learning biases, which may be genetic or culturally acquired, are studied through lab experiments and field observations. For example, in the lab it has been found that people transmit information about social relations more faithfully than non-social information, and field observations have revealed that pregnant women in certain Fijian villages acquire adaptive food taboos from prestigious, unrelated older women. Feeding these types of data into the cultural-population models produces population dynamics that are very different from those described by models assuming that cultural traits are genetically or even epigenetically transmitted (reviewed by Mesoudi, 2016).

Current studies of non-selective and selective social learning in non-human animals are based on field observations made using video and recording technologies, interventions such as 'seeding' new behaviours into populations, cross-fostering experiments, and transmission chain experiments in which information is transmitted along chains of individuals, as in the game of Chinese whispers. Many species have been found to have cultural traditions involving song dialects, food preferences, habitat preference, foraging methods, migration patterns or social conventions and to learn these behaviours from knowledgeable or successful adults. Incorporating the short- and long-term effects of social learning can provide evolutionary interpretations of animal behaviours that differ from conventional explanations. An example is the observed conflicts between parents and offspring, which, rather than being interpreted as the outcome of genetic conflicts, can be seen as resistance by and punishment of the young during the limited period in which they must learn basic survival skills (Avital and Jablonka, 2000).

Patterns of socially learned behaviours are sometimes related to changes in the genetic composition of populations, although the relation need not be causal. Hal Whitehead (2017) observed that in matrilineal killer whale species, where daughters remain with their mothers for life and sons live within the maternal pod but mate outside it, the genetic diversity of mitochondrial DNA is very low.

He suggests that this is the result of what he calls 'cultural hitchhiking': when a particular tradition becomes established by selection and drift (the pods consist of about 10 individuals), the maternal mitochondrial DNA, which, like the behaviour, is transmitted by the females, is inevitably correlated with ('hitchhikes' on) the tradition.

3.8 The Evolutionary Effects of Symbol-Based Learning

Like cultural evolution in whales and birds, human cultural evolution can lead to developmental changes and drive genetic evolution by constructing persistent ecological, social and cognitive developmental niches that impose new selective regimes on individuals and groups. The domestication of cattle and spread of dairy farming, a cultural process, may have driven the spread of the genes allowing lactose absorption; overcrowding in European cities seems to have led to the evolution of resistance to multiple infectious diseases; deforestation led to the spread of malaria and drove the selection of alleles in the haemoglobin coding genes that helped people cope with the malarial parasite; and the spread of increased copy number of the amylase gene is correlated with increased dietary starch intake and may have followed altered dietary traditions in hominins (Laland, 2017). In all these cases, symbolic representation and communication were involved, although this is not reflected in evolutionary models; the symbolic, collective and normative aspects of the changed tradition are only implied.

One of the best current examples of a human-specific symbolic-cultural change that has driven a genetic change is that which has resulted in a doubling during the last 200 years in the frequency of people in the USA who are homozygous for defects in the connexin gene. Homozygosity for this gene is the most common cause of genetic deafness. The increased frequency of connexin deafness is thought to be the result of the introduction of sign language about 400 years ago and the later establishment of schools for the deaf. Sign language enabled deaf people, who had previously tended to be socially isolated and less likely to have children, to become more integrated into the community, so selection against deafness was relaxed. More importantly, once schools for the deaf had been established, the likelihood that deaf people would find a partner with whom they could communicate through sign language and who might also carry the same deafness gene grew, so their children had an increased chance of being homozygous (Arnos *et al.,* 2008).

The evolution of sign language itself probably involves cultural processes similar to those that drove the evolution of spoken language in our ancestors (Jablonka and Lamb, 2014). Cecilia Heyes (2018) maintains that cultural transmission rather than genetic change has been dominant in constructing human-specific cognitive mechanisms such as the linguistic capacity. She argues that learning through precise imitation, explicit mind-reading and language are all 'cognitive gadgets' – products of cumulative cultural evolution rather than genetically evolved adaptive cognitive strategies. However, innate cognitive biases can have substantial effects on human cultural evolution (Claidière et al., 2014). For example, languages' colour terms commonly follow a particular order – languages with only two terms have black/dark and white/bright terms, those with three colour terms also have red, and those with four have yellow or green as well. This probably reflects a genetic bias in colour perception. In addition to innate biases, ecological-geographical conditions have been important in many aspects of human cultural evolution during the last 12,000 years. A well-discussed example is the way the distribution of domesticable species has influenced the geographical patterns of agriculture, warring practices and cultural expansion (Diamond, 1997). These and other large-scale historical trends, such as the rise and fall of empires, can be analysed by population models that incorporate geography, social scale, economy, features of governance and information systems (Turchin et al., 2018). Changes in institutional structures, literacy and numeracy, for example, depend on symbolic communication and cultural-symbolic evolution.

A different approach, inspired by Waddington's epigenetic landscape model, can be used to interpret stability and change within communities. For example, sociological processes can lead to persistent cultural 'landscapes': the persistence of urban poverty in the USA is a case where social and epigenetic factors seem to interact and form a self-sustaining sociological-cultural loop. Those identified include the kinds of jobs available, parents' education, the quality of inner-city schools, the structure of the welfare system, the consumption of junk food and use of drugs by parents and children, which all affect people's psychological and physiological well-being, and the exclusion of poor relatives by those who 'make it' in society (Tavory, Ginsburg and Jablonka, 2014).

3.9 Integrating and Estimating the Effects of Inheritance Systems

Whichever trait is the focus of an evolutionary study, the variety of hereditary factors operating at different levels of organization means that an integrative approach is required for its analysis. Systems Theory, which analyses the overall structure and behaviour of a system rather than treating it as the sum of its components, seems to be the appropriate framework for such studies (Jablonka and Noble, 2019).

A causal, mechanism-oriented analysis of inclusive inheritance is best captured within the framework of Developmental Systems Theory (DST), which focuses on the construction and reconstruction of lifecycles using a variety of developmental resources. Heredity is seen as an aspect of development, and the origin of heritable variations and their transfer are therefore analysed as developmental processes (Oyama, Griffiths and Gray, 2001). DST is thus well suited to the analysis of complex heritable traits such as food preferences, parenting styles, and mate and habitat choice, which stem from genetic, epigenetic and behavioural networks and, in humans, the symbolic system. Once a trait is characterized, at whatever level, its frequency, its value and its change over time need to be measured. Its change in frequency will depend on its transmissibility, its response to selection, and the population size and structure.

More than 50 years ago, George Price developed an equation that is a useful general formulation for estimating the evolutionary change in a trait at any level of description. His preliminary account of this was published in 1970, but his arguments are more developed in a paper that was probably written in 1971 but remained unpublished until 1995 (Price, 1970, c.1971/1995). These papers describe the change between generations in the average value of a trait as a consequence of selection (the covariance of the value of the trait with reproductive success or its proxy) and transmissibility bias (the bias introduced by processes such as mutation, epimutation, recombination, environmental induction, developmental biases, etc.).[8] Because of its generality, the equation can incorporate any heritable input – genetic, epigenetic-cellular, physiological-organismal, behavioural-cultural, or a combination of all these. Therefore, the transmissibility of the trait and the covariance of the trait-value with fitness can be decomposed into genetic, epigenetic and cultural components (Day and Bonduriansky, 2011). Such decomposition can tell us about the heritable inputs that affect transmissibility and selection, and the equations can also describe the relations between them, which may be quite complex. For example, genetic and cultural inputs need not go in the same direction.

Price's general formulation is not the only formal approach that can be applied to both genetic and non-genetic inheritance and evolution. There are three others that can be used to estimate the effects of various heritable inputs on a trait's frequency. The first, which was referred to in Sections 3.2 and 3.7, is population models that, by incorporating cultural and epigenetic inheritance, extend and modify population-genetics models. The second is an extension of quantitative-genetics models. When non-genetic inheritance is considered within this framework, one needs to change the way in which the phenotypic

---

[8] The 'transmissibility bias' term is complex because it is entangled with developmental selection (see, for example, the discussion by Uller and Helanterä, 2017).

variance is partitioned. Heritable phenotypic variance is decomposed into distinct genetic, non-genetic (e.g. epigenetic or cultural) and niche-constructed environmental variances, which alter the relationship between environmental and genetic variance and, when combined, account for the overall variance of the trait (Danchin et al., 2011; Tal, Kisdi and Jablonka, 2010).

A third approach treats inheritance systems as special types of communication systems that lead to the transmission of information, which is measured as the number of decodable variant states in a source. Heritable information can be decomposed into variations that are transmitted through various different routes between senders (ancestors) and receivers (descendants). Since information is substrate neutral, different types of informational inputs can be added, subtracted and so on. To achieve this generality, a translation of extended evolutionary biology models into informational terms needs to be developed (see Section 4.5). Such models would have to accommodate the fact that selection and transmissibility are not independent variables. Selection regimes are the outcome of the activities of individuals and communities, with a variety of legacies being transmitted between generations.

On their own, none of the three approaches provides a complete account of hereditary transmission. They illuminate different aspects of inheritance, so they should be seen as complementary.

4 Philosophical Implications: Is an Extended Evolutionary Synthesis Necessary?

A broad perception of heredity has implications for every aspect of the philosophy of evolutionary biology and for the philosophy of biology in general. Here we look at just a few of the philosophical issues that are particularly affected by our heredity-oriented approach: the structure of evolutionary theory, and how we construe the distinction between replicator and interactor; proximate and ultimate causes; the concept of homology; the notion of biological individuality and identity; and the meaning of 'information' in inheritance and evolution. We end by examining the claim that the recognition of different inheritance systems, alongside other closely related and overlapping topics, requires an extension or replacement of the twentieth-century Modern Synthesis of evolution, and appraising the nature of this conceptual change.

4.1 The Structure of Evolutionary Theory: Is Evolution Darwinian?

There are broad and narrow notions of Darwinian evolution and of Darwin's selection theory, so the question of whether evolution as we understand it in the twenty-first century is Darwinian depends on how we define Darwinism and how broadly we apply the principle of selection. Many scholars have emphasized that *The Origin* described not a single monolithic theory but a bundle of theories, of which two were fundamental. The first is the theory of descent with modification, which is a general descriptive framework for biological evolution. It assumes that there are hereditary processes that relate descendants to ancestors but does not specify how inherited modifications are generated and transmitted. The second is Darwin's selection theory, which is focused on the evolution of complex adaptations. Darwin suggested that those heritable variants within a population that contribute to the success (survival and reproduction) of individuals become more common in the population; these variations can accumulate and result in increased organizational complexity. As discussed in Section 1, the neo-Darwinian version of Darwin's selection theory, which incorporated the assumption that the origin of heritable variation is always blind (so selection is the only direction-giving process in evolution), became identified with 'Darwinism'. Any theory that supplemented natural selection with other directional processes was labelled 'non-Darwinian'. As is clear from the well-known last paragraph of *The Origin*, Darwin's evolutionary theory was, in this neo-Darwinian sense, non-Darwinian.

In the twentieth century there were several influential formulations of Darwin's selection theory. A very general one, which adopted the MS assumptions about the obligatory blindness of variation, was given by psychologist Donald Campbell (1974). He proposed that blind variations that are selectively retained form the basis of all adaptations, whether genetic, individually learned or cultural. Biologists were more permissive: Lewontin (1970) suggested that Darwin's scheme includes three principles and that while they hold, a population will undergo evolutionary change:

1. Different individuals in a population have different morphologies, physiologies and behaviours (phenotypic variation).
2. Different phenotypes have different rates of survival and reproduction in different environments (differential fitness).
3. There is a correlation between parents and offspring in the contribution of each to future generations (fitness is heritable). (Lewontin, 1970, p. 1)

Another influential general formulation was made by Maynard Smith (1986, chapter 1), who highlighted heredity and made explicit the commonly held assumption that reproduction involves multiplication. According to Maynard Smith, evolution occurs when there is:

1. Multiplication – an entity can reproduce to give two or more others.
2. Variation – not all entities are identical.
3. Heredity – like begets like. If there are different types of entities in the world, the result of the multiplication of entity of type A will be more entities of type A, while the result of the multiplication of entity B will be more of type B.
4. Competition – some of the heritable variation affects the success of entities in surviving and multiplying.

In neither Lewontin's nor Maynard Smith's formulations are the origins of variations (whether blind or partially directed), the mode of heredity (whether replicative or reconstructive) or the level at which success or fitness applies specified. Nor does their focus on selection exclude other directional processes that could lead to cumulative effects (e.g. repeated induction or internal developmental biases). What *is* assumed, though – and this assumption is, as far as we can tell, accepted by all evolutionary biologists, including us – is that cumulative adaptive evolution requires selection processes. There can be no complex adaptation without selection at some level. However, this does not mean that we can explain all the observed rates, directions and patterns of adaptive evolution without invoking other processes that interact with selection. Some adaptations are inexplicable on the assumption of selection alone.

In the light of our discussion in Sections 2 and 3 about the properties of non-genetic inheritance systems and their interactions, how do we answer the question 'Is evolutionary theory Darwinian?' Our short answer is that it is Darwinian if we include (i) multiple processes for the generation of blind and semi-directed heritable variation (genetic, epigenetic, behavioural and symbolic); (ii) both replicative and reconstructive processes of heredity; (iii) various forms and levels of selection; and (iv) the plausibility of rapid (saltational) changes. Our notion of Darwinian evolution, illustrated in Figure 2, is based on broad, development-oriented notions of variation, heredity and selection. It is different from the MS version of Darwinism in that all ten of the excluded and marginalized assumptions outlined in Section 1 are included in our twenty-first-century Darwinian evolutionary synthesis.

A longer answer to the question 'Is evolutionary theory Darwinian?' requires some conceptual unpacking that is related to old and current discussions in the philosophy of biology. First, our approach requires a reconsideration of the distinctions and relationships between units of evolution, units of heredity, units of development and targets of selection. Within the neo-Darwinian, gene-centred framework formulated in the 1970s, the DNA-gene (the genetic replicator) is the unit of biological heredity, selection and evolution, while the developing and interacting individual (the 'interactor', as Hull, 1980, called

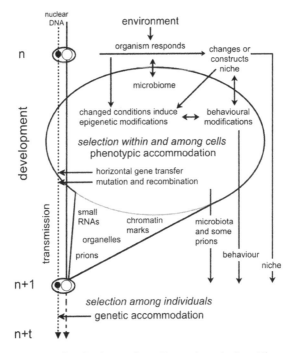

Figure 2 Processes impinging on heredity and evolution. The organism's responses at generation n are the outcome of its ancestral genetic, epigenetic, microbiotic, behavioural and ecological legacies, which interact with the environmental conditions in which it finds itself and lead to developmental selection and phenotypic accommodation. Changes in hereditary resources are transmitted to the next generation (n+1). They include variations in nuclear and cytoplasmic DNA that are the result of mutation, recombination and sometimes horizontal gene transfer; developmentally accommodated epigenetic factors; constructed niches; a modified microbiome; and behavioural adjustments. Variations are transferred to the next generation through both soma and germline and, if they persist, may lead to genetic accommodation. The ellipses at n and n+1 represent the germline, which includes nuclear DNA (black circle) and hereditary epigenetic factors in the nucleus (empty circle) and cytoplasm, which also contains organelle DNA.

it) is the unit of development. The individual (or the trait) is a target of selection, but on the assumption that genes have additive effects on fitness, selection can be assigned to the DNA-gene without conceptual loss. Similarly, although evolution happens in populations, so the population is the unit of evolution, evolutionary change is best traced by following changes in genes. Within this scheme, heredity is regarded as replicative (non-reconstructive), and the origin of variation is blind with respect to function (Dawkins, 1976, 1982).

From today's perspective, however, it seems that the gene is not the only replicator. Another unit that has been identified as having many gene-like attributes is the meme, and the selection of memes and memeplexes (networks of memes) is supposed to explain cultural evolution (Dawkins, 1976). Although selection is central to the meme-based view of cultural evolution, Daniel Dennett (2017), who describes memes as '*ways of doing things* that are trans-mitted nongenetically' (p. 224, his italics), points out that variations in these ways-of-doing arise during their development and reproduction, and in only some cases depend on human intelligent design. Kim Sterelny and his collea-gues (1996) have generalized the idea of the DNA replicator even further and applied it to methylation patterns, the three-dimensional structures of prions, physiological and behavioural patterns of activity, ideas and artefacts. However, although variations in these entities may involve blind, error-prone replicative processes, they also entail developmental reconstruction processes. Since all non-genetic 'replicators' are phenotypic traits that are products of development, the distinction between variable replicator and variable interactor is problematic.

Our alternative to the replicator/interactor distinction is to focus on heritably varying traits (HVTs) and reproducers, the entities that reproduce. An HVT is formed by a network of processes that is the outcome of one or more hereditary inputs. We believe that evolutionary analysis should start by following changes over time in the nature and frequency of HVTs, without privileging *a priori* any specific heritable input, such as a variation in DNA. An HVT can be a very low-level trait, like a pattern of chromatin, a prion or a self-sustaining loop, or a high-level trait, like a particular social network. Inputs into an HVT may include not only DNA variations but also heritable variations generated by other inheritance systems, so the relative significance of a particular input needs to be studied case by case. Because HVTs are units that develop and are reconstructed between generations, heredity and development are explicitly tied together (Jablonka, 2004b). The 'reproducer', a concept suggested by the American philosopher of biology James Griesemer (2000), is an alternative to the con-ventional target of selection, the individual. It is a unit of multiplication, development and hereditary variation. There is usually material overlap between parents and offspring, so parts of the parent entity are transferred to the offspring entity and confer developmental capacities on it, minimally the capacity to reproduce. A reproducer can be a replicating RNA molecule in the RNA world, a DNA methylation pattern, a cell, a multicellular organism, a social institution or a society. Evolution occurs in populations of reproducers.

The advantage of the HVT and reproducer concepts is that they allow us to think about heritable phenotypic variations at different levels of organization, about any mix of selective, inductive and niche-constructing processes, and about both replicative and constructive heredity. Focusing on HVTs and the processes that constitute and implement them allows a clearer distinction between targets of selection and units of evolution. It is an approach that is consistent with Doolittle's process-oriented view of evolution, which sees functional processes as units that are selected for their persistence and capacity for reproduction (Doolittle and Inkpen, 2018).

Discussion of whether or not evolution is Darwinian demands a shift away from the gene's-eye view and the recognition that selection is an important aspect of development. Selection can occur within the cell, among metabolic regulatory networks (as described by Braun for yeast – see Section 3.2), among dividing somatic cells and through phenotypic accommodation, which often depends on exploration and selective stabilization processes. When a developmentally selected variation is transmitted to the next generation of organisms, the effect is Lamarckian at the individual level, even when the generation of the variation depends on entirely blind processes at the intra-organismal level. Whether a process is regarded as 'Darwinian' or 'Lamarckian' therefore depends on the level of organization that is analysed. Many years ago, during the debates about the nature of bacterial adaptation, Peter Medawar cautioned against changing the level of analysis when comparing Darwinian and Lamarckian explanations, something that was and still is common when biologists try to avoid the spectre of Lamarckism. He pointed out that if you think about populations that gradually become resistant to penicillin (a seemingly Lamarckian transformation), it could be that within individual cells there are alternative pathways of metabolism that are mutually inhibitory, so that only one can prevail. Inhibiting one pathway means its replacement by another. Consequently,

> the Lamarckian transformation ... may be Darwinian at the lower analytical level represented by the enzymic population or complex of intersecting metabolic pathways within the individual bacterial cell. Such a description would be pointless for any except explanatory purposes, but it shows that no discussion of the rival interpretative powers of Darwinism and Lamarckism can have any useful outcome unless a certain analytical level is defined and adhered to. Hereafter we shall be concerned with individual organisms as analytical units, for it is only in this context that the rivalry is of any moment. (Medawar, 1957, p. 82).

Stochastic processes are fundamental to the physical world organisms inhabit, and the generation of variation will always have a stochastic, exploratory aspect. However, this fundamental stochasticity is harnessed and tamed by developmental and evolutionary processes that bias the range of heritable variations (as in the induction of some epigenetic variations) and select among variants during development. Hence, in addition to stochastic and blind processes of variation generation followed by selection, which are regarded as 'Darwinian', there are also 'Lamarckian' transformative processes (Jablonka and Lamb, 1995, 2014; Gissis and Jablonka, 2011).

4.2 Proximate and Ultimate Causation

Recognizing non-genetic inheritance and developmentally produced heritable variations has implications for the conceptualization of the distinction between proximate and ultimate causes, which has become identified with the distinction between physiological-developmental and evolutionary accounts of biological features. In an influential paper, Mayr described two approaches, those of functional biologists and evolutionary biologists:

> The functional biologist deals with all aspects of the decoding of the pro-grammed information contained in the DNA code of the fertilized zygote. The evolutionary biologist, on the other hand, is interested in the history of these codes of information and in the laws that control the changes of these codes from generation to generation. In other words, he is interested in the causes of these changes. (Mayr, 1961, p. 1502).

Proximate causes are investigated by physiologists, embryologists and cell biologists, and ultimate causes – those that generated the proximate mechanisms through evolutionary processes of natural selection – are studied by evolutionary biologists. The proximate–ultimate distinction can be construed as a descendant of the Aristotelian distinction between material and mechanistic ('moving') causes, on the one hand, and formal and final causes, on the other. Although Mayr focused on natural selection, ultimate causation as conceived by most neo-Darwinians is not, as Scott-Phillips and his colleagues (2011) have claimed, concerned solely with the fitness consequences of traits, but with what Mayr saw as the history of genetic programs – with evolutionary processes in general. No evolutionary biologist today would deny that factors such as drift and migration have played a major role in evolutionary diversification, so it is misleading to think of evolutionary history as solely the product of selection processes. Although drift alone cannot account for adaptations, the interaction between drift and selection can explain how otherwise inaccessible adaptive peaks can be reached.

Mayr's proximate–ultimate distinction reinforced the MS assumption that understanding development was unnecessary for evolutionary explanations.

Development had to do with proximate causes and the 'how' questions in biology – questions about how the genetic program is actualized – whereas evolution was concerned with ultimate causes – with 'why' or 'how come' a particular genetic program exists. But as many biologists realized in the 1960s while Mayr was hammering home the importance of the difference between proximate and ultimate causes, the distinction is not clear-cut and indeed, can be misleading. Reconstructive, developmentally guided, epigenetic and cultural inheritance processes are part of both development and evolution, and can be regarded as both proximate and ultimate causes. If we ask the question 'Why is X there?', an important part of the answer may be that X is there because of environmental induction, the effects of which were inherited and led to that particular variant of the trait becoming prevalent even though it is selectively neutral or even slightly deleterious.

Laland and his colleagues (2011) have emphasized that a common and significant aspect of the persistence of X may be the niche-constructing activities of organisms, which not only give direction to selection but also affect other evolutionary processes such as drift and migration, which in turn affect niche construction. Such bidirectional processes are the norm during evolution and have to be part of evolutionary-developmental analyses. Even if the focus is on selection, it has to be recognized that selection processes can be a crucial part of development, as in the immune system of animals, in somatic selection in plants, and in selective stabilization in the nervous system. When the outcomes of developmental selection are inherited between generations, they have direct evolutionary effects, so the process is *both* developmental (proximate) and evolutionary (ultimate).

4.3 An Extended Notion of Homology

The concept of homology was important in both pre-Darwinian and post-Darwinian thinking. In 1843, Richard Owen defined a homologue as 'the same organ in different animals under every variety of form and function'. He distinguished it from an analogue, 'a part or organ in one animal which has the same function as another part or organ in a different animal'. For Owen, analogous structures had similar functions, but homologous structures, identified through the similarity of the positions and relationships of equivalent parts, reflected something more fundamental – a shared organization, an archetype, the universal plan used for an organism's construction. The archetype concept provided pre-Darwinian biologists with the basis of a natural system for classifying the various forms found in nature.

In the post-Darwinian era, homologous structures were reinterpreted in evolutionary terms: they were not variations of an ideal archetype but the result of descent with modifications from a common ancestor. Differences in homologous structures were the adaptive outcome of natural selection in different lineages. Investigating structural homology was a key part of comparative morphology and studies of phylogenetic relations, the dominant areas of interest for evolutionary biologists in the late nineteenth and early twentieth centuries. For some, the similarities and differences in ontogeny were a crucial part of their investigations. But as Ray Lankester, Darwin, and many others realized,

> community in embryonic structure reveals community of descent; but dissimilarity in embryonic development does not prove discontinuity of descent, for in one of two groups the developmental stages may have been suppressed, or may have been so greatly modified through adaptation to new habits of life, as to be no longer recognisable. (Darwin, 1872, p. 396)

The insight that homology at one level of biological organization does not depend on or require homology at another level is important when we think about what is meant by homology today, when the concept is applied to far more than structural features. Homologies are now sought at every level from DNA sequences to behavioural traits, and the findings are used to establish evolutionary relationships and construct phylogenies. Our focus in this Element is on inheritance, and we have stressed that the relative importance of different hereditary inputs into the next generation can change over evolutionary time (see Sections 1.5 and 3.3). For example, an environmentally induced trait may eventually become genetically assimilated and much less dependent on external stimuli; or a trait that was initially maintained mainly through cultural inheritance may become more dependent on genetic inheritance or *vice versa*. If, by definition, we say that traits are homologous when they are inherited continuously in two different lineages with a single common ancestor, can similar traits that are inherited through different inheritance systems be regarded as homologous? Powell and Shea (2014) have convincingly argued that they can. It is not difficult to see why. For example, whereas intergroup violence in chimpanzees may be based mainly on a genetic propensity for this behaviour, in humans it may be based on cultural rather than genetic evolution. Nevertheless, the trait is homologous in chimps and humans because it was continuously inherited from a common ancestor.

Darwin and his contemporaries recognized that traits can be homologous despite divergence in developmental processes; early geneticists recognized that homologous traits often have different genetic underpinnings; and today we can see that homology is retained despite differences in the type of hereditary inputs. It takes a distant evolutionary viewpoint to discern through the changing metabolic, developmental and hereditary processes that contribute to similar traits those that have the taxonomic continuity that indicates homology.

4.4 Inheritance Systems, Identity and the Evolution of New Levels of Individuality

Studies of homology have shown that the evolutionary persistence of a phenotype does not depend on having the same material inputs (e.g. genes), on having the same developmental mechanisms or on having the same transmission systems. What persists in a lineage is a dynamic system of interactions, the components of which are constantly turning over. Today, as in the early twentieth century when the organicists were trying to steer a way between vitalism and the mechanistic explanations of the living world (see Section 1.5), a growing number of philosophers of biology are arguing that it is more realistic and a better guide for research to describe the living world in terms of dynamic, interacting, temporally extended hierarchies of processes rather than in terms of things or substances:

> Matters look quite different when we view the biological world as being organized not as a structural hierarchy of things, but as a dynamic hierarchy of processes, stabilized at different timescales. At no level in the biological hierarchy do we find entities with hard boundaries and a fixed repertoire of properties. Instead, both organisms and their parts are exquisitely regulated conglomerates of nested streams of matter and energy. (Dupré and Nicholson, 2018, p. 27)

How does recognizing this impinge on current discussions of biological individuality, a central topic in the philosophy of biology? According to this 'processual' view, biological entities (genes, chromosomes, cells, tissues, organisms, behaviours, lifecycles, etc.) have to be recognized as transient patterns of relative stability on different timescales. An individual organism was traditionally seen as a countable, more or less well-delineated living unit, but this common-sense notion of a biological individual has been challenged by recent studies showing that symbiosis is ubiquitous. Gilbert and Tauber regard what is usually seen as an individual as a porous and fuzzy 'holobiont', a composite organism and eco-system, whose metabolism is shared between the host and its microbiome, and whose boundaries are in flux.[9] The multigenomic holobiont, which incessantly constructs its developmental-ecological niche, problematizes, they claim, the very notion of biological individuality.

Though acknowledged to be somewhat nebulous, most philosophers of biology continue to use the term 'individual' and describe it from two major, non-exclusive perspectives – evolutionary and physiological. An 'evolutionary individual' is a systematic target of selection, a bearer of fitness that evolved

---

[9] More details about the views of Gilbert and Tauber, Clarke, Pradeu, Skillings and others can be found in *Biology and Philosophy,* Vol. 31(6), 2016, which is devoted to 'Biological Individuality'.

when independently reproducing individuals at one level of organization became parts of a new, higher-level unit (see discussion in Clarke, 2016). The paradigmatic evolutionary individuals (italicized) discussed by Maynard Smith and Szathmáry (1995) are *independent genes* that combined to form *chromosomes*, *independent prokaryotic cells* that combined to form *eukaryotes*, *independent eukaryotic cells* that combined to become *multicellular organisms*, and *multicellular organisms* that combined to form *colonial organisms*. For Maynard Smith and Szathmáry, the emergence of higher-level individuals constituted a 'major evolutionary transition'.

The alternative view of an individual, a 'physiological individual', is equivalent to what in common language we mean by an 'organism' – a metabolizing, developmentally and functionally integrated, self-maintaining and self-determining entity. Pradeu (2016) argues that the immune system plays a central role in shaping physiological individuality because it is systemic, constitutes a discrimination mechanism delineating the organism's boundaries, and is present in all living beings. In contrast, Moreno and Mossio (2015) argue that there are multiple processes that contribute to the individual's autonomy, and their relative importance changes as new integrating systems, such as the nervous system, evolve. In spite of such differences in characterization, evolutionary and physiological individuality can usually be assigned to the same entity (e.g. a unicellular organism). However, not all evolutionary individuals (for example, chromosomes) are organisms. Nor, according to Skillings, are all physiological individuals coherent units of selection: some symbiotic or ecological systems that have a great deal of functional integration are not reproducing Darwinian individuals, although the specific lineages that comprise their parts are.

Heredity and persistence are relevant to all notions of individuality and identity. A physiological individual is the product of dynamic interactions on several spatial and temporal scales, and its functional integrity depends on the inherited inputs it receives. Its identity through time and its spatial distinguishability from other individuals stem from a web of different types of identity at different levels of organization. All conventional physiological individuals (organisms) have genetic and epigenetic identities; all have an ecological identity involving ecological-developmental niche construction, which includes relations with symbionts and immune-system-based delineation; in addition, some (neural animals) have neural identity, some among these are endowed with subjective experiencing and have psychological identities, and a smaller number also have cultural identities, which in humans are symbolically constructed. Identity, a defining attribute of individuality, can thus be seen as a nested network of relations with lateral, top-down and bottom-up interactions among sub- and super-systems. Understanding how in any specific case these nested systems interact to form a dynamic self-maintaining entity must involve the study of the various processes of constructive and replicative inheritance that contribute to them at every level.

Like the physiological individual, the evolutionary individual cannot be fully understood without considering the role of all inheritance systems (Jablonka, 1994). We have argued that the evolutionary transitions to new levels of individuality that were discussed by Maynard Smith and Szathmáry all involved the evolution or novel recruitment of non-genetic inheritance systems (Jablonka and Lamb, 2006). For example, the recruitment and sophistication of epigenetic inheritance enabled the transition to differentiated and integrated multicellular organisms, and led to the evolution of developmental strategies that unite the individual and overcome the potential problems of selfish activity in its components (see Section 3.6). Similarly, the transition to social groups and social traditions in animals depended on the ability to pay attention to and learn from others, and language in humans depended on the construction of a symbolic communication system. Social learning and language are information channels that allow selection of behavioural patterns and strategies, and in some conditions unite and integrate the individual group sufficiently for selection among cultural groups to occur.

Not every major transition involved an increase in hierarchical, nested organization. Maynard Smith and Szathmáry (1995) pointed out that neither the transition from RNA as gene and enzyme to DNA-genes and protein-enzymes, nor the transition to symbolic language, involved an increase in nested hierarchy. In these cases, the basis of the transition to a new level of individuality was the emergence of a new information processing system (DNA; symbolic language). In fact, all the transitions Maynard Smith and Szathmáry listed were described in terms of changes in the way information is encoded, stored and utilized. When this general informational framework is adopted, it is clear that several major transitions to new types of individuals were overlooked by Maynard Smith and Szathmáry. We argued (Jablonka and Lamb, 2006) that the evolution of nervous systems should be considered a major transition, since it involved fundamental changes in the encoding, storage and use of information, which led to a new kind of individual, the neural individual. Ginsburg and Jablonka (2019) extended this argument, suggesting that there were additional major transitions during the evolution of neural organisms, including the transition to animals that have subjective experiencing.

4.5 The Meaning of Information in Inheritance and Evolution

In biology, 'information' is used mainly in the context of genetics and evolution, communication theory and cognitive neuroscience. In all cases, Shannon's correlational concept of information – that a signal carries information about a source if the state of the source can be predicted from the signal – is the basis of the discussion. A different way of presenting Shannon information is as a reduction in uncertainty, where uncertainty measures the number of states a system might be in. For example, a particular gene reduces uncertainty about the developing phenotype; a particular communication cue reduces uncertainty about the state of the sender; or a particular pattern of neural signalling reduces uncertainty about the response of a receiver-neuron or a muscle. In all these biological systems, the reduction in uncertainty is receiver dependent (Jablonka, 2002). The receiver must process (or 'interpret') the input received from the source, which may be either biotic or abiotic. Thus, a blind mole will not interpret dark clouds in the autumn sky as predictive of rain because it cannot see them, while an intelligent ape will. Similarly, an ultraviolet pattern on a flower is information for a bee but not for a rat. There is nothing necessary about the informational value of a particular source.

The interpretation systems of biological receivers are 'designed'; they have evolved to enable the interpretation of restricted types of information that have a large domain of organizational variation: transcribing many variant DNA sequences into RNAs; processing many patterns of electrical spikes; or processing many visual patterns within a narrow range of light wavelengths. The term 'pattern' is of central importance here. Whether associated with evolved or non-evolved sources, there must be a consistent relation between variation in some aspect of the organization of the source and the functional response of the receiver. When an input conforms to a type, it is recognized by the interpretation system of the organism, whether or not it leads to an adaptive response. Moreover, the notion of misinterpretation or error is inherent in the notion of biological information, since mistakes in the process of interpretation can lead to non-adaptive responses. Receiver-dependent information in living organisms has causal effects on receivers (Pocheville, Griffiths and Stotz, 2017), and often these effects are functional. Functional information is defined as any difference in the (external or internal) environment of a system that has made a systematic, causal difference to that system's goal-directed behaviour, which in an evolutionary context is fitness-promoting (Jablonka, 2002).

We can think about heredity in terms of information transmission because a receiver (e.g. a cell in a body, or an individual in a group) may also be a sender of information across generations. We have argued that for information to be inherited it is necessary that

> a receiver interprets (or processes) an informational input from a sender who
> was previously a receiver. When the processing by the receiver leads to the

reconstruction of the same or a slightly modified organization-state as that in the sender, and when variations in the sender's state lead to similar variations in the receiver, we can talk about the hereditary transmission of information. . . . Clearly, if the hereditary transmission of information is seen in this way, there is no need to assume that all hereditary variations and all evolution depend on DNA changes. (Jablonka and Lamb, 2006, p. 237)

Variation in patterns of chromatin, in small RNA profiles, and in socially and culturally learned patterns of behaviour in a parent can all reduce the uncertainty about the phenotype of the offspring and lead to parent–offspring similarity. Recognizing that there are different inheritance systems means that information processed by all systems that contribute to heredity must be integrated if phenotypic evolution is to be understood. Such integration requires a broad theoretical framework that can accommodate all different hereditary inputs under one formalism.

In Section 3.9 we discussed the Price equation as the basis for a general formulation of the evolutionary change in a trait. Since the equation includes both a selection term and a transmission term, it is well suited for describing the effects of different types of heritable informational inputs, which can affect selection, transmission or both. An appropriately extended Price equation, reframed in informational terms, could tease apart different kinds of heritable inputs whatever their mode of inheritance, and whatever the level of selection and transmission bias that is involved. But is it possible to 'translate' evolutionary change into informational terms? Donaldson-Matasci and colleagues (2010) found that in fluctuating environmental conditions, the fitness benefit of being able to detect and respond to a cue is exactly equal to the mutual information (a measure of how you can predict the value of one variable, X, from the value of another variable, Y) between the cue and the environment. They gave a simple example: if a system could be in any of six equiprobable states, and a cue reduces the possibilities to three of these, the cue provides a twofold reduction in uncertainty. However, their analysis of evolutionary adaptive change in terms of information is tailored to the specific case of fluctuating conditions, and, as they recognize, the expression of evolutionary change in informational terms needs to be generalized. If the Price equation could be expressed in informational terms, this would provide common ground for the analysis of hereditary information at any level of organization and for any system of inheritance.

4.6 Inheritance Systems and the EES

There are many other topics in theoretical biology and in the philosophy of biology that can be seen in a fresh light when non-genetic inheritance, with the systems view it entails, is taken on board. These include the relations between plasticity and evolvability (Lamm and Jablonka, 2008); the notion of species and the tree of life (discussed by O'Malley, 2014); the relation between mechanistic explanation and systems biology models, especially when analysing highly canalized or plastic traits (Baedke, 2018); the evolution of cognition, from cells to animals and from animals to humans (Moreno and Mossio, 2015); and general models of learning that can be applied to evolution at any level (Watson and Szathmáry, 2016). This, too, is only a partial list.

Recognizing multiple inheritance systems clearly has an impact on many facets of the philosophy of biology and orthodox evolutionary theory, so can the empirical findings and conceptual changes we have discussed be accommodated within the mid-twentieth-century MS? Do we really need a new evolutionary synthesis? Opinions on this vary, and this is reflected in the different names given to the emerging view, such as the 'Integrated Synthesis', the 'Inclusive Synthesis' and the 'Post-Modern Synthesis'. The most common description is the 'Extended Evolutionary Synthesis' (EES), which has been used in the title of several influential books and papers. In spite of having some reservations about this label, because we see the changes as more than a mere extension of the MS, we use EES in the following discussion, as we have throughout this Element.

Wray and his colleagues (2014) responded to the suggestion that a new evolutionary synthesis is needed by insisting (i) that the MS has already incorporated most of the new findings; (ii) that the list of subjects that need to be incorporated and given centre stage is an *ad hoc* list; and (iii) that the data coming from the study of non-genetic inheritance and especially epigenetics is not strong enough to warrant a change. In Sections 2 and 3 we provided ample evidence for the ubiquity of non-genetic inheritance and its role in evolution. But as we have shown in this Section, evolutionary biologists and philosophers have to do more than confront new, uncomfortable data (e.g. the many cases of epigenetic inheritance); they may also need to reframe some major concepts. In order to help decide whether the thinking behind the EES entails a different approach to evolutionary investigations, we will summarize the differences from the MS that we have discussed, and point to the scientific practices involved in the EES and the way that its ideas are disseminated in both professional (esoteric) circles and more generally among practitioners of other disciplines and laypersons (exoteric circles).

We noted in Section 1.4 that all 10 of the ideas that the MS marginalized or excluded are included in the EES. Positively expressed, these are:

- There is more to inheritance than genes.
- Soft inheritance is an empirical fact.

- The unit of evolution is rarely the single gene.
- Developmental plasticity is an important driver of evolution.
- Saltational evolution is plausible; generative physico-chemical processes and developmental biases are important for explaining rates of evolutionary change.
- Macroevolution involves processes such as hybridization and symbiogenesis, which are additional to those seen in microevolution.
- Niche construction and the transfer of ecological and developmental legacies can help explain patterns and rates of evolution.
- Adaptive evolution involves multiple and interacting levels and types of selection.
- Patterns of phylogeny are not overwhelmingly tree-like.

The entanglement of developmental plasticity, developmental bias, niche construction and multiple systems of inheritance suggested by the above list requires a move away from a gene-centred MS view of evolution to a development-centred view (Laland *et al.,* 2014, 2015). The MS's simplifying assumptions about replication, variation and selection have to be abandoned. We have to accept that the transgenerational transmission of information often involves reconstruction, not copying, and that what is reconstructed may be modified in a guided (non-random) manner. We also have to accept that the type of selection that is important in evolution is not just natural selection in populations of reproducing individuals; selection also occurs within individuals during their development, and this type of selection does not always require reproduction. The EES incorporates these views of variation and selection.

The developmental, process- and system-oriented perspective of the EES bears a clear resemblance to the organicists' view of heredity and evolution, something that has not escaped the attention of historians and philosophers of biology (e.g. Peterson, 2016). Waddington's epigenetic-systems perspective, with its focus on the role of canalization and plasticity and on the agency of the organism in evolution, is central to the EES. Both the organicists' and the EES framework are explicitly organism rather than gene centred, and their evolutionary analyses start with phenotypic (rather than genotypic) variation, which entails a developmental-system approach to heredity and evolution.

In both scientific (esoteric) and extra-scientific (exoteric) circles, the EES view of biology, particularly the incorporation of non-genetic inheritance, has led to methodological and social changes:

- There are new mathematical and simulation models that extend classical models in population and quantitative genetics. They incorporate multiple inheritance systems and their interactions, niche construction, and developmental constraints and affordances.

- New model organisms are being used. For example, clonal hermaphrodites or self-fertilizing organisms (such as *Arabidopsis* in plants and *Daphnia* in animals) and highly social organisms like cetaceans are used to investigate non-genetic inheritance.
- New techniques for studying different inherited inputs are being developed: for example, CRISPR-Cas methods for manipulating methylation patterns, and methods for detecting methylated and unmethylated cytosines in ancient DNA.
- New laboratories for the study of epigenetic inheritance and niche construction are being established, and the scientists directing them are becoming influential.
- There is growing recognition in other scientific disciplines of the medical, environmental and social importance of an extended view of heredity and evolution. This is reflected in an increase in interdisciplinary projects.
- New professional books that address specific aspects of the EES are being published (e.g. Gilbert and Epel, 2015; O'Malley, 2014; Shapiro, 2011) as well as new narratives of the history of biology that stress the historical roots of the EES in the organicist movement (e.g. Peterson, 2016). New journals exploring epigenetics (e.g. *Environmental Epigenetics*) and evo-devo (e.g. *EvoDevo*) have been established.
- Popular books, podcasts and YouTube videos that discuss aspects of the EES, especially epigenetic inheritance, have been produced. Much of the discussion is unrealistically hopeful, scientifically naïve and sometimes downright misleading. It is reminiscent of the hype around genetics in the early decades of the twentieth century and around the genome project in the late 1990s and early 2000s. The epigenetics hype calls for extra caution from scientists engaged in research and popularization of the EES's claims.

4.7 Does the EES Require a Change in Thought Style?

How should we describe the change in the way of thinking about evolution that has been taking place since just before the turn of the century? What is the significance of the constellation of conceptual changes, commitments, methodologies, institutional changes, and esoteric and exoteric dissemination-dynamics? The lack of historical perspective makes it very difficult to evaluate the significance of the new approach, since it is very much still in the making. It is not clear whether the EES is going to replace the MS or, as its opponents hope, fade away. We cannot prophesize, but we can consider some of the ideas that sociologists of science have about the introduction of new theoretical frameworks. Their analyses may enable us to assess how the EES fares in the broader social and cultural context.

The conglomeration of new ideas and practices associated with the EES seems to embody what the sociologist of science Ludwik Fleck described as a change in 'thought style'. A thought style is the product of what he called a 'thought collective' – a specific community of interacting people who produce and share knowledge about a scientific (or some other) topic. For Fleck,

> It [a thought style] is characterized by common features in the problems of interest to a thought collective, by the judgment which the thought collective considers evident, and by the methods which it applies as a means of cognition. The thought style may also be accompanied by a technical and literary style characteristic of the given system of knowledge. (Fleck, 1935/1979, p. 99)

Fleck's thought styles are not the same as the six 'styles of thinking' that historian Alistair Crombie (1988) identified. These methods of studying the natural world, which emerged and became stabilized over historical time, are mathematical, experimental, hypothetical-analogical, comparative, statistical and historical-evolutionary. Fleck's thought styles are more restricted than Crombie's and are attributes of particular, limited fields of inquiry (e.g. evolutionary biology). They include a mixture of different methodological approaches (e.g. for evolutionary biology: historical, statistical, comparative and experimental). Nevertheless, as Ian Hacking (1992) argued, each Crombie style (method), which Hacking reincarnates as a 'style of reasoning', has become a model of how to be 'objective' when studying a subject using that method. In this sense, there is a similarity between Fleck's sociological account of how we construct scientific 'facts' and Hacking's style-of-reasoning framework, which describes how new kinds of evidence and new types of explanations provide new criteria for objectivity and produce stable knowledge.

The EES also conforms to what Thomas Kuhn described as a 'system of thought', a 'paradigm' or a 'disciplinary matrix' (DM), a notion closely related to Fleck's thought style. In the postscript to the second edition of his influential book *The Structure of Scientific Revolutions* (1970), Kuhn described a DM as an assembly of explicit or implicit factors that determine the coherence of a scientific community, and discussed four main sorts of components. The first, 'symbolic generalizations', refers to the expressions used without dissent by group members, which can be cast in a logical form (like a formula) or as definitions of fundamental terms (e.g. simultaneity in physics and heredity in biology). The second type of component is the 'models and metaphors' used. Kuhn gives the example of molecules of gas described as tiny elastic billiard balls in random motion in classical physics. In the MS version of evolutionary theory, the equivalent is the description of hereditary variation as random changes in DNA and the population-genetics models based on this assumption – models and metaphors that are challenged by the EES. The third component is 'values', by which Kuhn means general normative values. He gives the example of the preference for quantitative predictions over qualitative ones in physics. Among EES proponents, the focus on phenotypic rather than genotypic evolution, and the centrality of canalization and plasticity, make qualitative descriptions the starting point of evolutionary analysis. The fourth factor is what Kuhn calls exemplars:

> One of the fundamental techniques by which the members of a group, whether an entire culture or a specialists' sub-community within it, learn to see the same things when confronted with the same stimuli is by being shown examples of situations that their predecessors in the group have already learned to see as like each other and as different from other sorts of situations. (Kuhn, 1970, pp. 193–4)

The exemplars are the case studies that members of the community present to students who are being encultured within the DM. In the MS version of evolutionary biology, the evolutionary change in the colouration of the peppered moth *Biston betularia* or the explanation of the patterns of distribution of sickle cell anaemia have been used as exemplars on the basis of which other evolutionary puzzles can be solved, while the EES's exemplars include hereditary changes over time in lines of self-fertilizing organisms like *Arabidopsis*, clonal organisms, inbred animals or highly social animals.

Kuhn suggested that a change to a new DM amounts to a gestalt switch, which follows the accumulation of anomalies within the prevalent DM. Though drastic, the change depends on processes of 'translation', which enable the professional community to accept the new framework and the conceptual and methodological commitments it entails. It is akin to the learning of a new language and often leads to profound misunderstandings and clashes between the old and new 'linguistic' communities. In the case of the EES, the 'new' language is one that was transiently 'dead' – the organicists' language, which, with the updates and modifications that new data and practices entail, is being 'revived'. It's an example of the spiralling process of scientific change that Lenin alluded to (see this Element, p.1).

The intellectual conflicts inherent in the process of change have been studied by the philosopher of science Menachem Fisch (2017), who focuses on normative values that innovators face when they have to decide which problem is worth studying and which solution and methodology are valuable. He stresses the importance of 'trading' with scholars from different esoteric and exoteric circles. Especially important for Fisch is the exposure to the normative critiques of other professionals (when deemed sincere and serious – something that requires an emotional aspect of trust), which enables an innovator to transcend the limits of purely subjective self-assessment. This is by no means a straightforward process. Fisch notes that the work of innovators who actively introduce new perspectives, especially their early work, manifests what he calls 'creative dithering', an ambivalent split between the old and new systems of thought. Within the EES context, one can see such creative dithering in the work of Mary Jane West-Eberhard, one of the most prominent advocates of the developmental, phenotype-first view of evolution, who largely avoided the implications of epigenetic inheritance. Similarly, many evo-devo practitioners, though focusing on gene networks and acknowledging the plausibility of large, viable, regulatory variations, pay little attention to non-genetic inheritance.

As Fleck and others have noted, the need for change is often difficult to digest, especially by the older generation, and the establishment of a new thought style frequently has to wait for the next generation of scientists. According to Fleck, changes in a scientific thought style are associated not only with new developments in the specific discipline but also with transformations in the sociological and cultural spheres. He stresses the role of institutions, methodologies and the general cultural values that are shared within both esoteric and exoteric circles. To understand a change in thought style (such as the change in evolutionary thinking), it is important to see it in the context of more widespread transformations in science and culture in recent decades. For biologists today, these include the focus on the analysis of 'big data', the ecological crisis, and political and cultural changes since the mid-1990s, when old political and ideological boundaries began to be dissolved and new ones erected.

What, then, is the nature of the change to an EES thought style? The status of the transition to the EES is debated, and not just by its opponents. There are different opinions within the EES community itself, with some regarding the EES as an important but continuous expansion of the MS, while others see it as a *bona fide* replacement, a revolutionary transition. Denis Noble (2017), one of the most radical proponents of the EES, has pointed out that whole areas in disciplines such as economics, sociology, political science and philosophy have adopted a gene-centred evolutionary perspective, and a change in outlook is going to have enormous implications that will affect our concept of humanity and our daily lives. There is growing awareness that psychological and physiological stresses have transgenerational effects, that human social status and mental and physical health are affected through all dimensions of heredity, and that we live in the midst of a complex web of social and ecological interactions that up to a point are robust and self-sustaining but can also tip. This means that our responsibility for the next generations and for the ecological balance on our planet, both as individuals and as collectives, is already getting general attention. These and similar studies, Noble argues, will lead in time to a new thought style, which is not merely an extension but a replacement of the MS-based view.

We find it very difficult to assess the significance and nature of the change in evolutionary thinking that we are in the midst of, but we can see similarities to another scientific change, one that occurred in the mid-twentieth century. What we are witnessing in biological evolutionary thought seems to be conceptually similar to what happened in psychology in the 1960s, when the behaviourists' thought style was replaced by that of the cognitivists. Behaviourists viewed psychological processes in terms of relations among input, output and reinforcement, black-boxing intervening computational cognitive processes, while the cognitivists focused on the computational processes. Similarly, MS advocates think of evolution in terms of the relations between genetic mutation (input), phenotypic change (output) and selection (reinforcement), black-boxing development and agency, while EES proponents focus on development and agency and see them, along with selection, as central to evolutionary theory, being not only the outcomes of evolution but also its drivers. However, whereas with the switch to cognitivism, behaviourism was rapidly displaced and ruthlessly belittled, the role of DNA inheritance remains fundamental to the EES, and the importance of the selection of genetic variations is not challenged. The transition to an EES thought style is therefore likely to be smoother and more gradual than that from behaviourism to cognitivism. Nevertheless, it seems to us that the consensus around the EES is growing for reasons similar to those that drove the cognitive revolution. The MS thought style is bound to persist for a while, both because it can absorb criticism by redefining its boundaries and because of the persistence of the intellectual aversion reflex to anything associated with the inheritance of acquired characters. The cognitive psychologist Amos Tversky was reported to have said: 'Theories are not refuted. They are embarrassed.' Today the MS does seem to be somewhat embarrassed, but history tells us that scientists can live with embarrassment for a remarkably long time.

References

Aramayo, R. and Selker, E. U. (2013). *Neurospora crassa*, a model system for epigenetics research. *Cold Spring Harbor Perspectives in Biology*, **5**(10), a017921. doi:10.1101/cshperspect.a017921

Arnos, K. S., Welch, K. O., Tekin, M. *et al.* (2008). A comparative analysis of the genetic epidemiology of deafness in the United States in two sets of pedigrees collected more than a century apart. *American Journal of Human Genetics*, **83**(2), 200–7. doi:10.1016/j .ajhg.2008.07.001

Avital, E. and Jablonka, E. (2000). *Animal Traditions: Behavioural Inheritance in Evolution*. Cambridge, UK: Cambridge University Press.

Baedke, J. (2018). *Above the Gene Beyond Biology: Towards a Philosophy of Epigenetics*. Pittsburgh, PA: University of Pittsburgh Press.

Bélteky, J., Agnvall, B., Bektic, L. *et al.* (2018). Epigenetics and early domestication: differences in hypothalamic DNA methylation between red junglefowl divergently selected for high or low fear of humans. *Genetics Selection Evolution* **50**(1),13. doi:10.1186/s12711-018-0384-z

Bondur
iansky, R. and Day, T. (2018). *Extended Heredity: a New Understanding of Inheritance and Evolution*. Princeton, NJ: Princeton University Press.

Bowler, P. J. (2003). *Evolution: the History of an Idea*, 3rd ed. Berkeley, CA: University of California Press.

Boyd, R. and Richerson, P. J. (1985). *Culture and the Evolutionary Process*. Chicago, IL: University of Chicago Press.

Braun, E. and David, L. (2011). The role of cellular plasticity in the evolution of regulatory novelty. In S. B. Gissis and E. Jablonka, eds., *Transformations of Lamarckism*. Cambridge, MA: MIT Press, pp. 181–91.

Bronfman, Z. Z., Ginsburg, S. and Jablonka, E. (2014). Shaping the learning curve: epigenetic dynamics in neural plasticity. *Frontiers in Integrative Neuroscience* **8**, 55. doi:10.3389/fnint.2014.00055

Campbell, D. T. (1974). Evolutionary epistemology. In P. A. Schilpp, ed., *The Philosophy of Karl R. Popper*. LaSalle, IL: Open Court, pp. 412–63.

Cavalli-Sforza, L. L. and Feldman, M. W. (1981). *Cultural Transmission and Evolution*. Princeton, NJ: Princeton University Press.

Chakrabortee, S., Byers, J. S., Jones, S. *et al.* (2016). Intrinsically disordered proteins drive emergence and inheritance of biological traits. *Cell* **167**(2), 369–81. doi:10.1016/j.cell.2016.09.017

Charlesworth, A. G., Seroussi, U. and Claycomb, J. M. (2019). Next-gen learning: the *C. elegans* approach. *Cell* **177**(7), 1674–6. doi:10.1016/j.cell.2019.05.039

Charlesworth, D., Barton, N. H. and Charlesworth, B. (2017). The sources of adaptive variation. *Proceedings of the Royal Society B* **284**(1855), 20162864. doi:10.1098/rspb.2016.2864

Ciabrelli, F., Comoglio, F., Fellous, S. *et al.* (2017). Stable Polycomb-dependent transgenerational inheritance of chromatin states in *Drosophila*. *Nature Genetics* **49**(6), 876–86. doi.10.1038/ng.3848

Claidière, N., Scott-Phillips, T. C. and Sperber, D. (2014). How Darwinian is cultural evolution? *Philosophical Transactions of the Royal Society B* **369** (1642), 20130368. doi:10.1098/rstb.2013.0368

Clarke, E. (2016). A levels-of-selection approach to evolutionary individuality. *Biology and Philosophy* **31**, 893–911. doi:10.1007/s10539-016-9540-4

Cortijo, S., Wardenaar, R., Colomé-Tatché, M. *et al.* (2014). Mapping the epigenetic basis of complex traits. *Science* **343**(6175), 1145–8. doi:10.1126/science.1248127

Crombie, A. C. (1988). Designed in the mind: Western visions of science, nature and humankind. *History of Science* **26**, 1–12. doi:10.1177/007327538802600101

Danchin, E., Charmantier, A., Champagne, F. A. *et al.* (2011). Beyond DNA: integrating inclusive inheritance into an extended theory of evolution. *Nature Reviews Genetics* **12**(7), 475–86. doi:10.1038/nrg3028

Darwin, C. (1872). *The Origin of Species*, 6th ed. London: Murray.

Dawkins, R. (1976). *The Selfish Gene*. Oxford: Oxford University Press.

Dawkins, R. (1982). *The Extended Phenotype: the Gene as the Unit of Selection*. Oxford: Freeman.

Day, T. and Bonduriansky, R. (2011). A unified approach to the evolutionary consequences of genetic and nongenetic inheritance. *American Naturalist* **178**(2), E18–E36. doi:10.1086/660911

Dennett, D. C. (2017). *From Bacteria to Bach and Back: the Evolution of Minds*. London: Allen Lane.

Diamond, J. (1997). *Guns, Germs, and Steel: the Fates of Human Societies*. New York, NY: Norton.

Dobzhansky, T. (1958). Species after Darwin. In S. A. Barnett, ed., *A Century of Darwin*. London: Heinemann, pp. 19–55.

Donaldson-Matasci, M. C., Bergstrom, C. T. and Lachmann, M. (2010). The fitness value of information. *Oikos* **119**(2), 219–30. doi:10.1111/j.1600-0706.2009.17781.x

Doolittle, W. F. and Inkpen, S. A. (2018). Processes and patterns of interaction as units of selection: an introduction to ITSNTS thinking. *Proceedings of the National Academy of Sciences USA* **115**(16), 4006–14. doi:10.1073/pnas.1722232115

Dor, D. (2015). *The Instruction of Imagination: Language as a Social Communication Technology*. New York, NY: Oxford University Press.

Dor, D. and Jablonka, E. (2010). Plasticity and canalization in the evolution of linguistic communication. In R. K. Larson, V. Déprez and H. Yamakido, eds., *The Evolution of Human Language: Biolinguistic Perspectives*. Cambridge, UK: Cambridge University Press. pp. 135–47.

Dunoyer, P., Melnyk, C., Molnar, A. *et al.* (2013). Plant mobile small RNAs. *Cold Spring Harbor Perspectives in Biology* **5**(7), 017897. doi:10.1101/cshperspect.a017897

Dupré, J. and Nicholson, D. J. (2018). A manifesto for a processual philosophy of biology. In D. J. Nicholson and J. Dupré, eds., *Everything Flows: Towards a Processual Philosophy of Biology*. Oxford: Oxford University Press, pp. 3–45.

Ephrussi, B. (1958). The cytoplasm and somatic cell variation. *Journal of Cellular and Comparative Physiology* **52**(Suppl. 1), 35–53. doi:10.1002/jcp.1030520405

Fisch, M. (2017). *Creatively Undecided. Toward a History and Philosophy of Scientific Agency*, Chicago, IL: University of Chicago Press.

Fleck, L. (1935/1979). *Genesis and Development of a Scientific Fact*, trans. F. Bradley and T. J. Trenn. Chicago, IL: Chicago University Press.

Gapp, K., Jawaid, A., Sarkies, P. *et al.* (2014). Implication of sperm RNAs in transgenerational inheritance of the effects of early trauma in mice. *Nature Neuroscience* **17**(5), 667–9. doi:10.1038/nn.3695

Gilbert, S. F. and Epel, D. (2015). *Ecological Developmental Biology: the Environmental Regulation of Development, Health, and Evolution*, 2nd ed. Sunderland, MA: Sinauer.

Ginsburg, S. and Jablonka, E. (2019). *The Evolution of the Sensitive Soul: Learning and the Origins of Consciousness*. Cambridge, MA: MIT Press.

Gissis, S. B. and Jablonka, E. (2011). *Transformations of Lamarckism: from Subtle Fluids to Molecular Biology*. Cambridge, MA: MIT Press.

Gokhman, D., Meshorer, E. and Carmel, L. (2016). Epigenetics: It's getting old. Past meets future in paleoepigenetics. *Trends in Ecology and Evolution* **31**(4), 290–300. doi:10.1016/j.tree.2016.01.010

Griesemer, J. (2000). The units of evolutionary transition. *Selection* **1**(1), 67–80. doi:10.1556/Select.1.2000.1-3.7

Hacking, I. (1992). 'Style' for historians and philosophers. *Studies in History and Philosophy of Science Part A* **23**(1), 1–20. doi:10.1016/0039-3681(92)90024-Z

Hardy, A. (1965). *The Living Stream*. London: Collins.

Hernday, A. D., Lohse, M. B., Fordyce, P. M. *et al.* (2013). Structure of the transcriptional network controlling white-opaque switching in *Candida albicans*. *Molecular Microbiology* **90**(1), 22–35. doi:10.1111/mmi12329

Heyes, C. (2018). *Cognitive Gadgets: the Cultural Evolution of Thinking*. Cambridge, MA: Harvard University Press.

Hodgson, S., de Cates, C., Hodgson, J. *et al.* (2014). Vertical transmission of fungal endophytes is widespread in forbs. *Ecology and Evolution* **4**(8), 1199–208. doi:10.1002/ece3.953

Holliday, R. (1987). The inheritance of epigenetic defects. *Science* **238**(4824), 163–70. doi:10.1126/science.3310230

Holliday, R. and Pugh, J. E. (1975). DNA modification mechanisms and gene activity during development. *Science* **187**(4173), 226–32. doi:10.1126/science.187.4173.226

Hu, J. and Barrett, R. D. H. (2017). Epigenetics in natural animal populations. *Journal of Evolutionary Biology* **30**(9), 1612–32. doi:10.1111/jeb.13130

Hull, D. (1980). Individuality and selection. *Annual Review of Ecology and Systematics* **11**, 311–32. doi:10.1146/annurev.es.11.110180.001523

Huxley, J. S. (1942). *Evolution, the Modern Synthesis*. London: Allen & Unwin.

Jablonka, E. (1994). Inheritance systems and the evolution of new levels of individuality. *Journal of Theoretical Biology* **170**(3), 301–9. doi:10.1006/jtbi.1994.1191

Jablonka, E. (2002). Information: its interpretation, its inheritance, and its sharing. *Philosophy of Science* **69**(4), 578–605. doi:10.1086/344621

Jablonka, E. (2004a). The evolution of the peculiarities of mammalian sex chromosomes: an epigenetic view. *BioEssays* **26**(12), 1327–32. doi:10.1002/bies.20140

Jablonka, E. (2004b). From replicators to heritably varying traits: the extended phenotype revisited. *Biology and Philosophy* **19**, 353–75. doi:10.1023/B:BIPH.0000036112.02199.7b

Jablonka, E. (2017). The evolutionary implications of epigenetic inheritance. *Interface Focus* **7**(5), 20160135. doi:10.1098/rsfs.2016.0135.

Jablonka, E. and Lamb, M. J. (1989). The inheritance of acquired epigenetic variations. *Journal of Theoretical Biology* **139**(1), 69–83. doi:10.1016/S0022-5193(89)80058-X

Jablonka, E. and Lamb, M. J. (1990). The evolution of heteromorphic sex chromosomes. *Biological Reviews* **65**(3), 249–76. doi:10.1111/j.1469-185X.1990.tb01426.x

Jablonka, E. and Lamb, M. J. (1995). *Epigenetic Inheritance and Evolution: the Lamarckian Dimension*. Oxford: Oxford University Press.

Jablonka, E. and Lamb, M. J. (2005). *Evolution in Four Dimensions: Genetic, Epigenetic, Behavioral and Symbolic Variations in the History of Life*, 1st ed. Cambridge, MA: MIT Press.

Jablonka, E. and Lamb, M. J. (2006). The evolution of information in the major transitions. *Journal of Theoretical Biology* **239**(2), 236–46. doi:10.1016/j.jtbi.2005.08.038

Jablonka, E. and Lamb. M. J. (2010). Transgenerational epigenetic inheritance. In M. Pigliucci and G. B. Müller, eds., *Evolution – the Extended Synthesis*. Cambridge, MA: MIT Press, pp. 137–74.

Jablonka, E. and Lamb, M. J. (2011). Changing thought styles: the concept of soft inheritance in the 20th century. In R. Egloff and J. Fehr, eds., *Vérité, Widerstand, Development: At Work with/Arbeiten mit/Travailler avec Ludwik Fleck*. Zürich: Collegium Helveticum, pp. 119–56.

Jablonka, E. and Lamb, M. J. (2014). *Evolution in Four Dimensions: Genetic, Epigenetic, Behavioral and Symbolic Variations in the History of Life*, 2nd ed. Cambridge, MA: MIT Press.

Jablonka, E. and Noble, D. (2019). Systemic integration of different inheritance systems. *Current Opinions in Systems Biology* **13**, 52–8. doi:10.1016/j.coisb.2018.10.002

Jablonka, E. and Raz, G. (2009). Transgenerational epigenetic inheritance: prevalence, mechanisms, and implications for the study of heredity and evolution. *Quarterly Review of Biology* **84**(2), 131–76. doi:10.1086/598822

Johannsen, W. (1911). The genotype conception of heredity. *American Naturalist* **45**(531), 129–59.

Klosin, A., Casas, E., Hidalgo-Carcedo, C. *et al.* (2017). Transgenerational transmission of environmental information in *C.* elegans. *Science* **356** (6335), 320–3. doi:10.1126/science.aah6412

Koonin, E. (2019). CRISPR: a new principle of genome engineering linked to conceptual shifts in evolutionary biology. *Biology and Philosophy* **34**(1), 9. doi:10.1007/s10539-018-9658-7

Kronholm, I., Bassett, A., Baulcombe, D. *et al.* (2017). Epigenetic and genetic contributions to adaptation in *Chlamydomonas. Molecular Biology and Evolution* **34**(9), 2285–306. doi:10.1093/molbev/msx166

Kuhn, T. S. (1970). *The Structure of Scientific Revolutions*, 2nd ed. Chicago, IL: University of Chicago Press.

Lachmann, M. and Jablonka, E. (1996). The inheritance of phenotypes: an adaptation to fluctuating environments, *Journal of Theoretical Biology* **181** (1), 1–9. doi:10.1006/jtbi.1996.0109

Laland, K. N. (2017). *Darwin's Unfinished Symphony: How Culture Made the Human Mind*. Princeton, NJ: Princeton University Press.

Laland, K. N., Sterelny, K., Odling-Smee, J. *et al.* (2011). Cause and effect in biology revisited: is Mayr's proximate-ultimate dichotomy still useful? *Science* **334**(6062), 1512–16. doi:10.1126/science.1210879

Laland, K., Uller, T., Feldman, M. *et al.* (2014). Does evolutionary theory need a rethink? Yes, urgently. *Nature* **514**(7521),161–4. doi:10.1038/514161a.

Laland, K. N., Uller, T., Feldman, M. W. *et al.* (2015). The extended evolutionary synthesis: its structure, assumptions and predictions. *Proceedings of the Royal Society B* **282**(1813), 20151019. doi:10.1098/rspb.2015.1019

Lamb, M. J. (2011) Attitudes to soft inheritance in Great Britain, 1930s–1970s. In S. B. Gissis and E. Jablonka, eds., *Transformations of Lamarckism*. Cambridge, MA: MIT Press, pp. 109–20.

Lamm, E. (2018). Inheritance systems. In E. N. Zalta, ed., *Stanford Encyclopedia of Philosophy* (Winter 2018 edn), Stanford, CA: Stanford University. https://plato.stanford.edu/archives/win2018/entries/inheritance-systems

Lamm, E. and Jablonka, E. (2008). The nurture of nature: hereditary plasticity in evolution. *Philosophical Psychology* **21**(3), 305–19. doi:10.1080/09515080802170093

Lenin, V. L. (1914/1930). *The Teachings of Karl Marx*. New York, NY: International Publishers.

Lewontin, D. (1970). The units of selection. *Annual Review of Ecology and Systematics* **1**, 1–18. doi:10.1146/annurev.es.01.110170.000245

Li, J., Browning, S., Mahal, S. P. *et al.* (2010). Darwinian evolution of prions in cell culture. *Science* **327**(5967), 869–72. doi:10.1126/science.1183218

Lindegren, C. C. (1966). *The Cold War in Biology*. Ann Arbor, MI: Planarian Press.

Logan, C. A. and Brauckmann, S. (2015). Controlling and culturing diversity: experimental zoology before World War II and Vienna's Biologische Versuchsanstalt. *Journal of Experimental Zoology* **323A**(4), 211–26. doi:10.1002/jez.1915

Markel, A. L. and Trut, L. N. (2011). Behavior, stress, and evolution in light of the Novosibirsk selection experiments. In S. B. Gissis and E. Jablonka, eds., *Transformations of Lamarckism*. Cambridge, MA: MIT Press, pp. 171–80.

Maynard Smith, J. (1966). *The Theory of Evolution*, 2nd ed. Harmondsworth, UK: Penguin.

Maynard Smith, J. (1986). *The Problems of Biology*. Oxford: Oxford University.

Maynard Smith, J. and Szathmáry, E. (1995). *The Major Transitions in Evolution*. Oxford: Freeman.

Mayr, E. (1961). Cause and effect in biology. *Science* **134**(3489),1501–6. doi:10.1126/science.134.3489.1501

Mayr, E. (1980). Prologue: some thoughts on the history of the evolutionary synthesis. In E. Mayr and W. B. Provine, eds., *The Evolutionary Synthesis: Perspectives on the Unification of Biology*. Cambridge, MA: Harvard University Press, pp. 1–48.

Mayr, E. (1982). The *Growth of Biological Thought: Diversity, Evolution, and Inheritance*. Cambridge, MA: Harvard University Press.

Mayr, E. and Provine, W. B. (1980). *The Evolutionary Synthesis: Perspectives on the Unification of Biology*. Cambridge, MA: Harvard University Press.

Medawar, P. (1957). *The Uniqueness of the Individual*. London: Methuen.

Mesoudi, A. (2016). Cultural evolution: a review of theory, findings and controversies. *Evolutionary Biology* **43**, 481–97. doi:10.1007/s11692-015-9320-0

Moreno, A. and Mossio, M. (2015). *Biological Autonomy: a Philosophical and Theoretical Enquiry*. Dordrecht: Springer.

Müller, G. B. (2017). *Vivarium. Experimental, Quantitative, and Theoretical Biology at Vienna's Biologische Versuchsanstalt*. Cambridge, MA: MIT Press.

Nanney, D. L. (1958). Epigenetic control systems. *Proceedings of the National Academy of Sciences USA* **44**(7), 712–17. doi:10.1073/pnas.44.7.712

Nätt, D., Rubin, C-J., Wright, D. *et al.* (2012). Heritable genome-wide variation of gene expression and promoter methylation between wild and domesticated chickens. *BMC Genomics* **13**, 59. doi:10.1186/1471-2164-13-59

Nicholson, D. J. and Dupré, J. (2018). *Everything Flows: towards a Processual Philosophy of Biology*. Oxford: Oxford University Press.

Noble, D. (2017). *Dance to the Tune of Life: Biological Relativity*. Cambridge, UK: Cambridge University Press.

O'Malley, M. A. (2014). *Philosophy of Microbiology*. Cambridge, UK: Cambridge University Press.

Owen, R. (1843). *Lectures on Comparative Anatomy Delivered at the Royal College of Surgeons in 1843*. London: Longman, Brown, Green, and Longmans.

Oyama, S., Griffiths, P. E. and Gray, R. D. (2001). *Cycles of Contingency*. Cambridge, MA: MIT Press.

Pál, C. (1998). Plasticity, memory and the adaptive landscape of the genotype. *Proceedings of the Royal Society B* **265**(1403), 1319–23. doi:10.1098/rspb.1998.0436

Peterson, E. L. (2016). *The Life Organic: the Theoretical Biology Club and the Roots of Epigenetics*. Pittsburgh, PA: University of Pittsburgh Press.

Pocheville, A. and Danchin, E. (2017). Genetic assimilation and the paradox of blind variation. In P. Huneman and D. Walsh, eds., *Challenging the Modern Synthesis: Adaptation, Development, and Inheritance*. New York, NY: Oxford University Press, pp. 111–36.

Pocheville, A., Griffiths, P. E. and Stotz, K. (2017). Comparing causes: an information-theoretic approach to specificity, proportionality and stability. In H. Leitgeb, I. Niiniluoto, P. Seppälä *et al.*, eds., *Proceedings of the Fifteenth Congress of Logic, Methodology and Philosophy of Science*. London: College Publications, pp. 261–86.

Powell, R. and Shea, N. (2014). Homology across inheritance systems. *Biology and Philosophy* **29**, 781–806. doi:10.1007/s10539-014-9433-3

Pradeau, T. (2016). Organisms or biological individuals? Combining physiological and evolutionary individuality. *Biology and Philosophy* **31**, 797–817.

Price, G. R. (1970). Selection and covariance. *Nature* **227**, 520–1. doi:10.1038/227520a0

Price, G. R. (c.1971/1995). The nature of selection. *Journal of Theoretical Biology* **175**(3), 389–96. doi:10.1006/jtbi.1995.0149

Provine, W. B. (2001). *The Origins of Theoretical Population Genetics, with a New Afterword*. Chicago, IL: University of Chicago Press.

Przibram, H. (1903). Die neue Anstalt für experimentelle Biologie in Wien. *Verhandlungen der Gesellschaft deutscher Naturforscher und Ärzte* **74**, 152–5.

Przibram, H. (1912). Die Umwelt des Keimplasmas. 1. Das Arbeitsprogramm. *Archiv für Entwicklungsmechanik* **33**, 666–81.

Quadrana, L. and Colot, V. (2016). Plant transgenerational epigenetics. *Annual Review of Genetics* **50**, 467–91. doi:10.1146/annurev-genet-120215-035254

Rechavi, O. (2014). 'Guest list' or 'black list'? Heritable small RNAs as immunogenic memories. *Trends in Cell Biology* **24**(4), 212–20. doi:10.1016/j.tcb.2013.10.003

Richards, C. L., Alonso, C., Becker, C. *et al.* (2017). Ecological plant epigenetics: evidence from model and non-model species, and the way forward. *Ecology Letters* **20**(12), 1576–90. doi:10.1111/ele.12858

Rigal, M., Becker, C., Pélissier, T. *et al.* (2016). Epigenome confrontation triggers immediate reprogramming of DNA methylation and transposon silencing in *Arabidopsis thaliana* F1 epihybrids. *Proceedings of the*

National Academy of Sciences USA **113**(14), E2083–E2092. doi:10.1073/pnas.1600672113

Riggs, A. D. (1975). X inactivation, differentiation, and DNA methylation. *Cytogenetics and Cell Genetics* **14**(1), 9–25.

Rivoire, O. and Leibler, S. (2014). A model for the generation and transmission of variations in evolution. *Proceedings of the National Academy of Sciences USA* **111**(19), E1940–E1949. doi:10.1073/pnas.1323901111

Rodrigues, J. A. and Zilberman, D. (2019). Evolution and function of genomic imprinting in plants. *Genes and Development* **29**(24), 2517–31. doi:10.1101/gad.269902.115

Sager, R. and Kitchin, R. (1975). Selective silencing of eukaryotic DNA. *Science* **189**(4201), 426–33. doi:10.1126/science.189.4201.426

Schmalhausen, I. I. (1949). *Factors of Evolution: the Theory of Stabilizing Selection*, trans. I. Dordick. Philadelphia, PA: Blakiston.

Scott-Phillips, T. C., Dickins, T. E. and West, S. A. (2011). Evolutionary theory and the ultimate–proximate distinction in the human behavioral sciences. *Perspectives on Psychological Science* **6**(1), 38–47. doi:10.1177/1745691610393528

Shapiro, J. A. (2011). *Evolution: a View from the 21st Century*. Upper Saddle River, NJ: FT Press Science.

Simpson, G. G. (1953). The Baldwin effect. *Evolution* **7**(2), 110–17.

Skinner, M. K., Gurerrero-Bosagna, C., Haque, M. M. *et al.* (2014). Epigenetics and the evolution of Darwin's finches. *Genome Biology and Evolution* **6**(8), 1972–89. doi:10.1093/gbe/evu158

Smith, T. A., Martin, M. D., Nguyen, M. *et al.* (2016). Epigenetic divergence as a potential first step in darter speciation. *Molecular Ecology* **25**(8), 1883–94. doi:10.1111/mec.13561

Smocovitis, V. B. (1996). *Unifying Biology. The Evolutionary Synthesis and Evolutionary Biology*. Princeton, NJ: Princeton University Press.

Soen, Y. (2014). Environmental disruption of host–microbe co-adaptation as a potential driving force in evolution. *Frontiers in Genetics* **5**, 168. doi:10.3389/fgene.2014.00168

Soen, Y, Knafo, M and Elgart, M. (2015). A principle of organization which facilitates broad Lamarckian-like adaptations by improvisation. *Biology Direct* **10**, 68. doi:10.1186/s13062-015-0097-y

Soto, C. (2012). Transmissible proteins: expanding the prion heresy. *Cell* **149** (5), 968–77. doi:10.1016/j.cell.2012.05.007

Sperber, D. (1996). *Explaining Culture: a Naturalistic Approach*. Oxford: Blackwell.

Stajic, D., Perfeito, L. and Jansen, L. E. T. (2019). Epigenetic gene silencing alters the mechanisms and rate of evolutionary adaptation. *Nature Ecology and Evolution* **3**(3), 491–8. doi:10.1038/s41559-018-0781-2

Sterelny, K., Smith, K. C. and Dickison, M. (1996). The extended replicator. *Biology and Philosophy* **11**(3), 377–403. doi:10.1007/BF00128788

Tal, O., Kisdi, E. and Jablonka, E. (2010). Epigenetic contribution to covariance between relatives. *Genetics* **184**(4), 1037–50. doi:10.1534/genetics.109.112466

Tavory, I., Ginsburg, S. and Jablonka, E. (2014). The reproduction of the social: a developmental system approach. In L. R. Caporael, J. R. Griesemer and W. C. Wimsatt, eds., *Developing Scaffolds in Evolution, Culture, and Cognition*. Cambridge, MA: MIT Press, pp. 307–25.

Tikhodeyev, O. N. (2018). The mechanisms of epigenetic inheritance: how diverse are they? *Biological Reviews* **93**(4), 1987–2005. doi:10.1111/brv.12429

Tomasello, M. (2014). *A Natural History of Human Thinking*. Cambridge, MA: Harvard University Press.

Turchin, P., Currie, T. E., Whitehouse, H. *et al.* (2018). Quantitative historical analysis uncovers a single dimension of complexity that structures global variation in human social organization. *Proceedings of the National Academy of Sciences USA* **115**(2), E144–E151. doi:10.1073/pnas.1708800115

Uller, T. and Helanterä, H. (2017). Heredity and evolutionary theory. In P. Huneman and D. Walsh, eds., *Challenging the Modern Synthesis*. New York: Oxford University Press, pp. 280–316.

Van der Graaf, A., Wardenaar, R., Neumann, D. A. *et al.* (2015). Rate, spectrum, and evolutionary dynamics of spontaneous epimutations. *Proceedings of the National Academy of Sciences USA* **112**(21), 6676–81. doi:10.1073/pnas.1424254112

Vilcinskas, A. (2016). The role of epigenetics in host–parasite coevolution: lessons from the model host insects *Galleria mellonella* and *Tribolium castaneum*. *Zoology* **119**(4), 273–280. doi:10.1016/j.zool.2016.05.004

Vrana, P. B. (2007). Genomic imprinting as a mechanism of reproductive isolation in mammals. *Journal of Mammalogy* **88**(1), 5–23. doi:10.1644/06-MAMM-S-013R1.1

Waddington, C. H. (1957). *The Strategy of the Genes*. London: Allen & Unwin.

Waddington, C. H. (1960). Evolutionary adaptation. In S. Tax, ed., *Evolution after Darwin, Vol. 1. The Evolution of Life*. Chicago, IL: University of Chicago Press, pp. 381–402.

Waddington, C. H. (1975). *The Evolution of an Evolutionist*. Edinburgh: Edinburgh University Press.

Watson, R. A. and Szathmáry, E. (2016). How can evolution learn? *Trends in Ecology and Evolution* **31**(2), 147–57. doi:10.1016/j.tree.2015.11.009

Weber, B. H. and Depew, D. J. (2003). *Evolution and Learning: the Baldwin Effect Reconsidered.* Cambridge, MA: MIT Press.

Weismann, A. (1889). *Essays upon Heredity and Kindred Biological Problems.* Vol. 1. Trans. and ed. E. B. Poulton, S. Schönland and A. E. Shipley. Oxford: Clarendon Press.

Weissman, C. (2011). Germinal selection: a Weismannian solution to Lamarckian problematics. In S. B. Gissis and E. Jablonka, eds., *Transformations of Lamarckism.* Cambridge, MA: MIT Press, pp. 57–66.

Wells, H. G., Huxley, J. S. and Wells, G. P. (1929–1930/1934). *The Science of Life.* 3 vols. London: Amalgamated Press.

West-Eberhard, M. J. (2003). *Developmental Plasticity and Evolution.* New York, NY: Oxford University Press.

Whitehead, H. (2017). Gene–culture coevolution in whales and dolphins. *Proceedings of the National Academy of Sciences USA* **114**(30), 7814–21. doi:10.1073/pnas.1620736114

Whitehead, H. and Rendell, L. (2014). *The Cultural Lives of Whales and Dolphins.* Chicago, IL: University of Chicago Press.

Williams, G. C. (1966). *Adaptation and Natural Selection.* Princeton, NJ: Princeton University Press.

Wray, G. A, Hoekstra, H. E., Futuyma, D. J. *et al.* (2014). Does evolutionary theory need a rethink? No, all is well. *Nature* **514**(7521), 161–4. doi:10.1038/514161a

Zhang, T-Y. and Meaney, M. (2010). Epigenetics and the environmental regulation of the genome and its function. *Annual Review of Psychology* **61**, 439–66. doi:10.1146/annurev.psych.60.110707.163625

Zheng, X., Chen, L., Xia, H. *et al.* (2017). Transgenerational epimutations induced by multi-generation drought imposition mediate rice plant's adaptation to drought condition. *Scientific Reports* **7**, 39843. doi:10.1038/srep39843

Cambridge Elements ☰

Philosophy of Biology

Grant Ramsey
KU Leuven
Grant Ramsey is a BOFZAP research professor at the Institute of Philosophy, KU Leuven, Belgium. His work centres on philosophical problems at the foundation of evolutionary biology. He has been awarded the Popper Prize twice for his work in this area. He also publishes in the philosophy of animal behaviour, human nature and the moral emotions. He runs the Ramsey Lab (theramseylab.org), a highly collaborative research group focused on issues in the philosophy of the life sciences.

Michael Ruse
Florida State University
Michael Ruse is the Lucyle T. Werkmeister Professor of Philosophy and the Director of the Program in the History and Philosophy of Science at Florida State University. He is Professor Emeritus at the University of Guelph, in Ontario, Canada. He is a former Guggenheim fellow and Gifford lecturer. He is the author or editor of over 60 books, most recently *Darwinism as Religion: What Literature Tells Us about Evolution*; *On Purpose*; *The Problem of War: Darwinism, Christianity, and their Battle to Understand Human Conflict*; and *A Meaning to Life.*

About the Series
This Cambridge Elements series provides concise and structured introductions to all of the central topics in the philosophy of biology. Contributors to the series are cutting-edge researchers who offer balanced, comprehensive coverage of multiple perspectives, while also developing new ideas and arguments from a unique viewpoint.

Cambridge Elements ≡

Philosophy of Biology

Printed in the United States
By Bookmasters